The
Psych
101
Series

James C. Kaufman, PhD, Series Editor
Neag School of Education
University of Connecticut

Erin N. Colbert-White, PhD, is an associate professor of psychology at the University of Puget Sound in Tacoma, Washington, where she teaches courses such as Introductory Psychology, History and Systems of Psychology, and Learning and Behavior. She holds a doctorate in psychology from the University of Georgia. Her graduate work was a case study exploring an African grey parrot's pragmatic use of speech during interactions with her human caregiver. Dr. Colbert-White's areas of interest are the history of the field of animal cognition, animal consciousness, origins of language, and social behavior. She has worked with a variety of species, including bottlenose dolphins, capuchin monkeys, and Harris's hawks. Dr. Colbert-White devotes much of her time to developing her skills as a teacher and undergraduate mentor.

Allison B. Kaufman, PhD, is a research scientist with the Department of Ecology and Evolutionary Biology at the University of Connecticut and an adjunct professor on the University of Connecticut's Avery Point Campus. She holds a doctorate in neuroscience from the University of California, Riverside. Dr. Kaufman has published on language and cognition in several species of nonhuman animals and is the coeditor of three other books: *Animal Creativity and Innovation*, *Pseudoscience: The Conspiracy Against Science*, and *The Scientific Foundation of Zoos and Aquariums: Their Role in Conservation and Research*. She is involved in the zoo/aquarium community as both a researcher and an animal care volunteer.

Animal Cognition 101

Erin N. Colbert-White, PhD
Allison B. Kaufman, PhD

SPRINGER PUBLISHING COMPANY
NEW YORK

Copyright © 2020 Springer Publishing Company, LLC

All rights reserved.

No part of this publication may be reproduced, stored in a retrieval system, or transmitted in any form or by any means, electronic, mechanical, photocopying, recording, or otherwise, without the prior permission of Springer Publishing Company, LLC, or authorization through payment of the appropriate fees to the Copyright Clearance Center, Inc., 222 Rosewood Drive, Danvers, MA 01923, 978-750-8400, fax 978-646-8600, info@copyright.com or on the Web at www.copyright.com.

Springer Publishing Company, LLC
11 West 42nd Street
New York, NY 10036
www.springerpub.com
http://connect.springerpub.com

Acquisitions Editor: Rhonda Dearborn
Compositor: Amnet Systems

ISBN: 978-0-8261-6234-2
e-book ISBN: 978-0-8261-6235-9
DOI: 10.1891/9780826162359

19 20 21 22 / 5 4 3 2 1

The author and the publisher of this Work have made every effort to use sources believed to be reliable to provide information that is accurate and compatible with the standards generally accepted at the time of publication. The author and publisher shall not be liable for any special, consequential, or exemplary damages resulting, in whole or in part, from the readers' use of, or reliance on, the information contained in this book. The publisher has no responsibility for the persistence or accuracy of URLs for external or third-party Internet websites referred to in this publication and does not guarantee that any content on such websites is, or will remain, accurate or appropriate.

Library of Congress Cataloging-in-Publication Data
Names: Colbert-White, Erin N., author.
Title: Animal cognition 101 / Erin N. Colbert-White, PhD, Allison B. Kaufman, PhD.
Other titles: Animal cognition one hundred and one
Description: New York, NY : Springer Publishing Company, LLC, [2020] | Includes bibliographical references and index.
Identifiers: LCCN 2019010559 (print) | LCCN 2019011280 (ebook) | ISBN 9780826162359 (eBook) | ISBN 9780826162342 (print : alk. paper)
Subjects: LCSH: Cognition in animals.
Classification: LCC QL785 (ebook) | LCC QL785 .C534 2020 (print) | DDC 591.5/13—dc23
LC record available at https://lccn.loc.gov/2019010559

> Contact us to receive discount rates on bulk purchases.
> We can also customize our books to meet your needs.
> For more information, please contact: sales@springerpub.com

Erin N. Colbert-White: https://orcid.org/0000-0001-9952-9016
Allison B. Kaufman: http://orcid.org/0000-0003-1511-4927

Publisher's Note: New and used products purchased from third-party sellers are not guaranteed for quality, authenticity, or access to any included digital components.

Printed in the United States of America.

For Cody Brooks, who fostered my passion and
made me the open-minded skeptic I am today.
I hope you think this book is "not bad."
—*ECW*

For James, who never expected to carry around
a hairless rat in his shirt pocket, teach a bird to
talk like Hannibal Lecter, trip over four animals in
one morning (every morning), or
have 500 crickets in his laundry room.
—*ABK*

Contents

Preface ix

Chapter 1 Historical Perspective on Animal Cognition 1

Chapter 2 Theoretical and Methodological Approaches to Animal Cognition 31

Chapter 3 Consciousness in Animals 67

Chapter 4 Communication Between Animals 103

Chapter 5 Social Cognition in Animals 141

Chapter 6 Cognitive Flexibility in Animals 177

Chapter 7 Individual Differences Between Animals 215

Chapter 8 Conclusion 247

Index 257

Preface

veryone loves animals. We learn about them in zoos and aquariums, rehabilitate them when they're sick, observe their habits and abilities, and treat them as members of our families. Animals empathize with us, are happy to see us when we come home, make mischief, play games, solve problems, and stash treats for later.

Or do they? Animals are often romanticized, and books that describe their behavior can appeal more to a reader's heart than mind. In reality, researchers who study animal behavior are very practical; there is far more room for science than romance. That is where this book comes in. By introducing the history, essential theories and methodology, major topics, research findings, and controversies in the field of animal cognition, we aim to present an accessible, critical, and scientific look at the field of animal cognition. This brief overview offers what it is that scientists do when they study animal cognition, as well as an informed account of what we know (or at least what we think we know) about the cognitive abilities of other animals.

INTENDED AUDIENCE

By approaching this book with an eye toward encouraging readers to think critically about animal headlines they might see in the media, our goal is to challenge myths and preconceived notions. We accomplish this partly by including explanations of the scientific method, the perils of anthropomorphism, and the importance of distinguishing between correlation and causation. This orientation makes *Animal Cognition 101* a fantastic introductory read for anyone who is interested in learning the science behind animal cognition. Further, the inclusion of methods makes the book a useful supplement for advanced high school psychology as well as lower-division undergraduate courses. College professors teaching cognitive psychology, for example, may find it particularly helpful as a complement to their human cognitive texts.

As researchers, both of us have been heavily trained in biology and evolutionary theory. This means if someone asked us to name the "smartest" animal, our response would be, "It depends. How are you measuring 'smart'?" By this, we mean that it's important to evaluate animals' cognitive abilities within the context of the species being tested and the methods being used. Furthermore, the field of animal cognition overlaps with wide-reaching disciplines, including ethology (the study of how animals interact with and in their environment), evolutionary biology, and sociobiology (biological foundations for social behavior). Our backgrounds in evolutionary theory, as well as the interdisciplinary nature of the field, make *Animal Cognition 101* an appropriate supplementary text for a number of biology courses.

DISTINGUISHING FEATURES

One theme that we intentionally wove throughout the book is the importance of knowing a species' natural history before making assumptions or drawing conclusions about an animal's behavior. This means learning as much as possible about its biology, social organization, habitat, diet, predators, and other factors that might contribute to *why* a species might—from an evolutionary stance—have a particular cognitive ability. This helps draw better and more informed conclusions about data and avoids the trap of human bias. For example, one bias many people hold is that a bigger brain or a mammal brain or an insert-cute-animal-you-spend-a-lot-of-time-with brain means "more cognitive." As a result, people tend to write off fish as much less cognitive than what science has actually revealed. Considering there are more fish kept as pets than any other animal and the fact that we farm them by the billions as a food source, there's plenty of reasons to take the time to learn about their amazing cognitive abilities. To this end, you'll find that the collection of research findings in this book reflects a breadth of diversity, from insects to fish, to showcase more than just the charismatic megafauna we're most familiar with.

All seven content chapters include an "Animal Spotlight" and "Human Application" section. The Animal Spotlights highlight individual animals that have made significant contributions to an area relevant to the content discussed in each chapter. You may have heard of many of these animals. They are the "rock stars" of animal cognition, the ones who have made guest appearances on late-night television and got to hang out with well-known public figures like Mr. Rogers. The Human Application sections serve as a reminder that even though we use the word "animal" to imply "nonhuman animal" in this book, in reality, humans are animals, too. What we discuss in this book applies directly to human life. We'll use these Human Application sections to connect topics in animal cognition to human behavior and cognition and see how you can apply what you've learned to make a real difference in your own life.

Learning happens best when it is immersive and dynamic. At the ends of Chapters 2 through 7, you'll find an idea for how you can assume the role of a scientist and conduct some animal-friendly research of your own. These experiential ministudies are designed to be an extension of the content in each chapter and are a reflection of our desire to get readers excited about the science of animal cognition. Keeping in mind that not everyone has an animal in their home, access to nature parks, or readily available research materials, we developed the procedures to be as flexible and inclusive as possible for all readers. We hope that by both reading and doing, you'll gain a well-rounded introduction to the field of animal cognition.

CONTENTS

Before we get started, a few definitions are in order. The first has to do with the name of this book. Depending upon someone's training, focus, or generation, you may notice that the name for our field varies. Originally, it was pioneered as comparative psychology ("comparative" being in reference to humans), and those who later honed in specifically on cognitive abilities rather than sensory systems, for example, referred to the field as comparative cognition. Still, in other circles, the word "comparative" is dropped, and the study of animals'

PREFACE

cognitive abilities is simply called "animal cognition." For the purposes of this book, we'll be using "animal cognition."

When we say an animal is cognitive, or engaging in cognition, what does that mean? *Intelligence* might seem to suffice, but the study of animal intelligence is actually a subcategory of animal cognition research. Throughout the book, when we say "cognition," we'll have renowned animal cognition researcher Sara Shettleworth's (2001) definition in mind. According to her, cognition is "all ways in which animals take in information through the senses, process, retain and decide to act on it" (p. 277). An animal who is behaving cognitively, therefore, is one who is not behaving instinctually in its environment. A cognitive animal dynamically communicates, thinks, reflects, reacts, and is aware. While cognition inevitably manifests differently from species to species, there is a common underlying set of processes that are inherently active.

Animal Cognition 101 is divided into one history chapter, one theories and methods chapter, five topical chapters, and a final chapter on future directions. Each topical chapter addresses an overarching area of research in animal cognition, which we discuss across multiple species. In addition, we pay special attention to describing the different ways that researchers set up their studies to arrive at their conclusions. As is the general practice in the field, we'll be using the term "animal" to mean "nonhuman animal," and scientific names for species will be given the first time they are mentioned.

Chapter 1, Historical Perspective on Animal Cognition, is what a professor once called "ancestor worship." When did we decide to formally study how animals think, and how did we go about doing it? The chapter addresses some of the initial challenges the field faced, of which there were many, and some which remain unresolved to this day. The animal profiled is Kanzi the bonobo (*Pan paniscus*), perhaps one of the most well-known and groundbreaking animals in the history of animal cognition. To pay homage to Charles Darwin's idea that we can learn about ourselves by studying other animals, the Human Application section explores animals' involvement in biomedical research.

Within Chapter 1, we also take time to highlight two of the field's unsung heroes, Margaret Floy Washburn and Charles Henry Turner. Due to misogyny and racism, opportunities for both of these pioneers were limited, and yet they went on to be more prolific than many of their better-known white male counterparts. By sharing their stories, we hope to communicate two major ideas. First, science, for all its objectivity and quest for truth, does not operate on a meritocracy. Instead, the social climate largely dictates whose voices are heard and, therefore, who gets to contribute to what is known about how the world works. Second, for every one Turner and one Washburn, there are countless others who were and are unable to pursue their passion for science. Given Turner's and Washburn's enormous contributions to animal cognition, how might greater equity and inclusion at the turn of the 20th century have transformed the directions of the field? While we will never know, as you read this book, we ask that you keep this question in the back (or forefront) of your mind.

With all the Internet videos out there boasting everything from sledding crows to genius lizards, it was important to us to present a scientific representation of animals' cognitive abilities. For this reason, Chapter 2, Theoretical and Methodological Approaches to Animal Cognition, will orient you to what makes studying animal cognition a science (rather than a hobby), the major research designs researchers use, where research is conducted, how to think critically about intelligent animal videos that seem too good to be true, as well as what it means to collect data and how. Chapter 2 also spends a lot of time addressing the roles of anecdotes and anthropomorphism, neither of which are scientific, but which are relevant to animal cognition work. To illustrate both of these phenomena, a horse (*Equus caballus*) by the name of Clever Hans is the subject of the chapter's Animal Spotlight. Clever Hans seemed to many researchers—and for quite some time—to be able to count and do mathematics. The Human Application section in this chapter touches on the question of why we humans should care about animal cognition. It provides a brief introduction to tools used by evolutionary biologists[1] to determine how distantly related species are from one another and from humans. Using that information can inform us about how studying animal minds can teach us about the uniqueness—or lack thereof—of many human cognitive abilities.

Following the chapters on history and methodology, our goal is for readers to be prepared to read and evaluate the research findings of five major areas of animal cognition. The five topical chapters are designed to promote critical thinking by giving a background

---

1. We do mean brief, and we would like to politely request the evolutionary biologists reading this not to chase us with sticks.

in relevant terms, methodology, ways to interpret results, and controversies. For example, the main test for self-awareness, the mirror test, has been passed by ants but not parrots, which are renowned for their intelligence. We intentionally included surprising findings such as these throughout the book in an effort to challenge assumptions and biases readers might have about cognition in the animal kingdom. To that end, it was important to us as authors to not cherry-pick research findings that fit expectations, and to illustrate the gray areas in the field. Even in an introductory book, knowing the history, the controversies, and the criticisms are key to understanding the field.

Chapter 3, Consciousness in Animals, jumps right into the hotly debated area of animal consciousness. We take an in-depth look at how philosophers and scientists have defined consciousness, specific cognitive abilities that might signal consciousness, and which animals can be said to have them, or a version of them. The main topics covered include theory of mind, self-awareness, and emotions. Happy, the first elephant (*Elephas maximus*) documented to behave as if she recognized herself in a mirror, as well as the important implications of this finding, is the subject of the Animal Spotlight. The Human Application section will walk you through how theory of mind develops in children and the ways developmental psychologists can determine whether a child has mastered it.

Chapter 4, Communication Between Animals, focuses on communication, a topic that those who follow animal cognition research may be already familiar with. It addresses many different ways that animals communicate with each other, including vocal, gestural, and olfactory. Chapter 4 also distinguishes between research aimed at learning about animals' natural communication systems and humans' attempts to teach language-based communication systems to animals. The animals profiled here are a pair of bottlenose dolphins (*Tursiops truncatus*), Phoenix and Akeakamai, who participated in some of the first and most thorough studies of linguistic abilities in marine mammals. To show how vital language exposure is for children to develop it themselves, the Human Application chronicles the case study of Genie, a young girl who was found to have been deprived of almost all language exposure for her first 13 years of life.

Social cognition is featured in Chapter 5, Social Cognition in Animals. Social cognition involves the many complex ways in which animals engage socially among themselves. From imitation to cultural traditions passed down from generation to generation, there are many examples of how animals learn from one another. Chapter 5 also discusses empathy and social referencing, both of which play a part in an animal's ability to understand its place in a social group or situation. In the Animal Spotlight, we highlight Imo the Japanese macaque (*Macaca fuscata*). Imo's unique innovation to clean and season her food spread like wildfire throughout her group in a classic example of cultural transmission of behavior. The Human Application section introduces operant conditioning, a fundamental way organisms learn, to explain TAGteach, a popular movement that uses a "tag" in the form of a click or other sound to mark correct behavior. This method has been extremely successful in a variety of unexpected human-learning settings.

Chapter 6, Cognitive Flexibility in Animals, addresses the overall flexibility of the animal mind. For centuries, there have been those who believe animals are mindless behaving machines. You probably do not think that, or you would not be reading this book, but where is the line between instinct and cognitive behavior? Do animals plan out their actions in advance, play, and create? One creative crow (*Corvus moneduloides*) named Betty, and her remarkable ability to problem solve and use tools, is featured in the Animal Spotlight. In the Human Application section, we'll tackle the difficult question of how to measure creativity in humans and offer tips for how you can find and increase creativity and innovation in your own life.

Finally, Chapter 7, Individual Differences Between Animals, reminds us that despite the fact that research findings teach us what species on the whole *can* do, not all animals within a species are the same; individual differences exist. Using constructs like personality, emotional well-being, and intelligence, Chapter 7 explains how researchers account for and measure dispositional and performance-based differences in animals. Here the spotlight shines on the African grey parrot (*Psittacus erithacus*) Alex, a one-of-a-kind individual who remains unparalleled in what he taught us about the abilities of these not-so-bird-brained animals. Personality tests are explored in the Human Application sections. How do the ones you're doing on social media really measure up to those that have been developed by psychologists?

The final chapter of the book, Chapter 8, Conclusion, brings everything together. Here, we offer the major take-home messages that we hope you've learned, as well as a collection of ideas for where the field may be heading as it begins its second century. One of these directions is a new branch of research that asks us to think with

PREFACE

greater urgency not just about how humans impact animals' lives via habitat destruction and climate change, but about how our rapid change on the planet has affected animals' minds as well.

As you'll soon discover, it's not simply enough to be entertained or awestruck by the remarkable feats of the animal mind, though we, like you, are continually amazed. Instead, we ask you to join us on a journey of thinking, reflecting, critiquing, and applying what has been learned so far about cognition in the animal kingdom.

REFERENCE

Shettleworth, S. J. (2001). Animal cognition and animal behavior. *Animal Behaviour, 61*, 277–286. doi:10.1006/anbe.2000.1606

ACKNOWLEDGMENTS

We are grateful to the editor and publisher—James C. Kaufman and Springer Publishing Company—of the Psych 101 Series for allowing us the privilege of contributing to their informative series. In particular, we are very thankful for the support of Mindy Chen and Rhonda Dearborn at Springer Publishing Company. Lastly, we would like to thank Tim Beyer, Steve Brown, Dominic Byrd, Jen Elwood, Adrian Hatfield, Barbara Lukas, Kristen Magoun, Anna Marchand, Amanda Nasser, Sandhy Reddy, Gwen Remington, Dylan Richmond, May Tang, Genet Tulgetske, Peter Wimberger, and Roberta Wright for their thoughts, support, and cheerleading throughout the publication process.

Genius 101
Dean Keith Simonton, PhD

IQ Testing 101
Alan S. Kaufman, PhD

Leadership 101
Michael D. Mumford, PhD

Anxiety 101
Moshe Zeidner, PhD
Gerald Matthews, PhD

Psycholinguistics 101
H. Wind Cowles, PhD

Humor 101
Mitch Earleywine, PhD

Obesity 101
Lauren M. Rossen, PhD, MS
Eric A. Rossen, PhD

Emotional Intelligence 101
Gerald Matthews, PhD
Moshe Zeidner, PhD
Richard D. Roberts, PhD

Personality 101
Gorkan Ahmetoglu, PhD
Tomas Chamorro-Premuzic, PhD

Giftedness 101
Linda Kreger Silverman, PhD

Evolutionary Psychology 101
Glenn Geher, PhD

Psychology of Love 101
Karin Sternberg, PhD

Intelligence 101
Jonathan Plucker, PhD
Amber Esping, PhD

Depression 101
C. Emily Durbin, PhD

History of Psychology 101
David C. Devonis, PhD

Psychology of Trauma 101
Lesia M. Ruglass, PhD
Kathleen Kendall-Tackett, PhD, IBCLC, FAPA

Memory 101
James Michael Lampinen, PhD
Denise R. Beike, PhD

Media Psychology 101
Christopher J. Ferguson, PhD

Positive Psychology 101
Philip C. Watkins, PhD

Psychology of Aging 101
Robert Youdin, PhD

Creativity 101, Second Edition
James C. Kaufman, PhD

Animal Cognition 101
Erin N. Colbert-White, PhD
Allison B. Kaufman, PhD

Historical Perspective on Animal Cognition

s pioneering psychologist Herman Ebbinghaus (1908) put it, "psychology has a long past, but a short history" (p. 3). What he meant by this was that even though humans have wondered about their own behavior and mental processes for thousands of years, the formal discipline of psychology was not established until much later—1879 to be exact. Similarly, the debate of whether animals simply behave or whether their behavior is informed by some kind of mind is also many centuries' old, but it was not until the early 20th century that the field of animal cognition came to be.

The sankofa bird is a common symbol in West African art that depicts a bird looking back over its shoulder. The translation of *sankofa* reflects the importance of looking back to the past. In this chapter, we explore the history of the field of animal cognition in order to help situate the current methods, theories, and interpretations of findings. After all, how can you truly understand the present field of animal cognition, or even speculate about its future, without being knowledgeable of how the field came to be? We trace two key figures, René Descartes and Charles Darwin, whose starkly different ways of viewing animal minds—driven by their own understanding of the natural world at the time—divided the scientific community for many generations. Out of Darwin's presumptions of the relationship between humans and animals, we highlight how the field of animal cognition took root, complete with a few residual growing pains left over from Descartes' perspective, and refined itself into a formal discipline that was heavily influenced by scientific and societal views of its time.

EARLY CONSIDERATIONS AND DEBATES

Descartes

Humans' fascination with animal behavior and cognition likely dates back as far as our fascination about our own behavior and mental processes. For the sake of brevity, we begin our historical journey in the 16th century with the Scientific Revolution. During this era of major scientific advancement, philosophers of the 16th and 17th centuries began studying animal behavior in a more systematic way than previous generations. Many of their questions focused on anatomy and structure with the ultimate goal of better understanding the human body. For example, physician William Harvey (1578–1657) experimented on the circulatory systems of both warm- and cold-blooded animals in order to "illustrate man by the structure of animals" (Power, 1898).

René Descartes (1596–1650) was one of Harvey's contemporaries who also compared human and animal behavior, cognition, and anatomy. Inspiration struck one day while Descartes was visiting a royal garden. At the time, a new fad for the French elite were water gardens containing *automata*, hydro-powered lifelike machines. The moving metal creatures became Descartes' muse, and he returned home to create one of the most well-known theories in modern philosophy. Descartes conceptualized the body to be like a machine; when stimulation was applied to the body, such as a child touching a candle flame, the brain responded by releasing animal spirits. This fluid traveled down channels, similar to the pipes inside the automata, and caused an involuntary reflexive movement of the hand away from the flame. No thoughts or fancy mental power is involved, just a basic reflexive motion system.

CHAPTER 1 HISTORICAL PERSPECTIVE ON ANIMAL COGNITION

Due to the strong infusion of Christianity into the Western culture at the time, including science and philosophy, Descartes drew a hard line in the sand between humans and animals. Though both had machinelike bodies that were governed by animal spirits, God had gifted humans with a *soul* (i.e., mind). This metaphysical soul within a machine body was the foundation for Descartes' iconic mind–body dualism perspective. This soul was located deep in the brain and controlled the rest of the body. Among other abilities, Descartes thought the soul allowed humans to reason; engage in voluntary, nonreflexive behaviors; use language to communicate; and think mathematically and abstractly. As we will discover, animal cognition research has contradicted the uniqueness of many of these abilities, but at the time there was no clear indication that animals were more than reflexive, unfeeling automata. Descartes' stance, backed up by the Christian belief of human superiority and dominion over animals, laid the foundation of how society and science came to view and treat animals for more than two centuries.

Before we get too far, a bit of a public service announcement is in order. When evaluating historical events and actions, it's important to consider the historical context within which they occurred. For example, vivisection, the act of performing invasive experimental surgeries on a live animal, was popular and accepted among physicians like William Harvey. While this may seem inhumane or barbaric by today's standards, the widespread understanding during Harvey's time was that animals were basically just machines, lacking both feeling and emotions. Knowing this, is it fair to judge someone for vivisecting in 1650 using today's standards? Probably not, especially given everything we learned about the human body from their work. Armed with a more informed understanding of the knowledge and values of Harvey's time, we can look back at his experiments and appreciate them for what they contributed to science—even if we do not condone them.

There are, of course, always exceptions. While the viewing of animals as machines and the practice of vivisection were common in the 16th to 19th centuries, some held strong dissenting opinions. Philosopher and animal advocate Voltaire (1694–1778) accused his medical contemporaries of hypocrisy, saying:

> Some barbarians seize this dog . . . they nail him to a table and dissect him living to show the mesenteric veins. You discover in him the same organs of sentiment which are

> in yourself. Answer me, machinist, has nature arranged all the springs of sentiment in this animal that he should not feel? Has he nerves, and is he incapable of suffering? (1901/1764)

This excerpt from Voltaire's "beasts" entry in his *Dictionnaire Philosophique* was published nearly a century before arguably the most influential book in animal cognition's history by Charles Darwin, *On the Origin of Species*. Yet, in Voltaire's writings, we already see inklings of what was to come. Voltaire and others poked holes in Descartes' line in the sand: If animals are similar enough to humans to allow for anatomical and physiological comparison, similarities between humans and nonhumans may not end with the physical body. If only Voltaire had been born a century later, he would have lived to see the first major shift toward an acknowledgment of animal minds.

Darwin's Continuity of Species

At the time of its publication in 1859, *On the Origin of Species* stirred international controversy, and understandably so. Armed with thousands of compelling observations, Charles Darwin (1809–1882) proposed nature had not been created in its current form by God, as society and even science at that time believed (Gould, 2002). Instead, the many species Darwin had observed seemed to better fit within a theory that they were the product of evolution over time in response to changes in their environment and competition for resources. As prominent zoologist Stephen J. Gould (2002) wrote, Darwin wanted society to stop viewing nature as some beautiful, majestic harmony and instead to see it as a "struggle for personal success" (p. 121), with animals and plants savagely competing with each other to survive and pass on their genes to the next generation. As if this was not enough, Darwin (1871) later published *The Descent of Man and Selection in Relation to Sex*, which made the mind-blowing conclusion that millions of years ago, humans shared a common ancestor with the other great apes. By kicking humans off their pedestal and inserting them into the animal kingdom, Darwin boldly redefined the centuries-old relationship between humans and animals—and made a lot of people really mad in the process. While these ideas were accepted among some academics at the time, as you might imagine, they were largely received as serious threats to Western Christian ideology.

Along with noticing a clear continuity of species with respect to similar anatomy and physiology among humans and animals, Darwin also noted continuity for psychological processes as well (e.g., emotions, concept formation, complex communication; Boakes, 1984). His *The Expression of the Emotions in Man and Animal*, for example, opened the door to thinking more seriously about the hypocrisy raised by Voltaire a century earlier. In one chapter, Darwin (1872) identified behavioral, vocal, and physiological similarities in how basic emotions such as terror, pain, and joy are experienced by both humans and animals. Using systematic (i.e., structured, organized) observations as evidence to support his theories, Darwin encouraged an entirely new way of viewing animals—one that cemented an interest in investigating animals' subjective experiences and their cognitive abilities.

Naturalists and Psychologists

Inspired by Darwin's findings, naturalists and early psychologists became increasingly curious about looking for parallels between humans' and animals' cognitive abilities. As the name might imply, naturalists are interested in observing the natural world, without intervention or experimentation. It is from naturalists that we come to understand the natural history of a species. Natural history includes observations of an animal's social, reproductive, habitational, morphological (i.e., how they look physically), cognitive, and foraging characteristics, as well as other features of that species in the wild. Darwin was a naturalist; we know this because he spent decades observing, recording, and sampling in order to develop his theory of how organisms change over time. By studying features of finches' natural history, for example, he was able to solve the mystery of their differently sized beaks—where you live dictates what's on the menu (e.g., seeds vs. nuts), and what's on the menu drives the evolutionary refinement of your food extraction tool (e.g., pointy vs. blunt beak).

Once word spread that animals might also have subjective experiences and cognitive abilities, many of the naturalists of Darwin's time set out to investigate his claims. One famous naturalist and friend of Darwin's was George Romanes (1848–1894). In his book *Animal Intelligence*, Romanes (1882) reported dozens of stories of complex cognitive behaviors in everything from scorpions to elephants. In those stories, he tended to overascribe human traits to animals, such as his account of scorpions choosing to commit suicide rather than burning in a fire or foxes vindictively getting back at their fox friends for failing a hunting expedition. This sort of 1:1 comparison of animals to humans, while a well-intentioned effort to "raise the brutes and lower the humans," as the trend was called, has many drawbacks that did not go unnoticed by his contemporaries.

While British naturalists laid the foundation in the late 19th century by reporting natural observations and stories about animals, American psychologists at the beginning of the 20th century drew conclusions about animals' cognitive abilities by turning to the laboratory. At the time, psychology was locked in a bitter dispute about the direction the field should take. Some, like Edward Titchener (1867–1927), argued for more subjective methods, such as introspection, to better understand the human mind. Introspection was a technique made popular by a psychology school of thought called structuralism. As its name implies, structuralists like Titchener were curious about the structure of the mind, that is, how is the mind organized, how does it process information, and so on. For example, Titchener might present rose oil to a research participant and ask them how it makes them feel or what they think about when they smell it. While it might seem like a good idea, introspection's subjectivity made it difficult to learn anything concrete about the mind, which frustrated psychologists like John Watson (1878–1958), who was tired of the subjectivity and lack of rigor his colleagues were using to study the human mind. In what is known by many as his Behaviorist Manifesto, Watson (1913) heavily criticized his fellow psychologists, calling for an end to studying any behavior that could not be outwardly observed and measured—including the mind. By avoiding studying something as abstract and unobservable as the mind, Watson thought psychology could join the ranks of sciences like biology and chemistry.

At this point, you might be asking, "How could a psychologist demanding that psychology *not* study mental processes contribute to a field that is all about studying animals' mental processes?" Well, it was not Watson's bitter rejection of psychology's "soft science" direction that mattered. Rather, the school of thought Watson created to replace structuralism, called behaviorism, established a new lens through which American psychologists would study cognitive abilities in humans and animals. For better or worse, behaviorism had a lasting effect on the field of animal cognition.

Along with Watson's appreciation (though not complete agreement; Logue, 1978) of Darwin's continuity of species argument, his call for greater objectivity and scientific rigor inspired the establishment of more and more animal learning and behavior laboratories across the United States. In these laboratories, psychologists avoided using people's animal stories as evidence, favoring instead to run formal experiments. Some researchers performed studies to gain a better understanding of an animal's subjective experience by providing different visual stimuli, for example, and measuring its reaction. This is how we learned that bees can see in color, for example. Other researchers manipulated features of the animal's environment or physiological state (e.g., restrict food to make the animal hungry) and recorded differences in the animal's behavior. Those differences in outward behavior, they argued, were observable evidence of what was going on inside the animal. For example, Tolman (1948) famously tested hungry rats in mazes where there was either food or no food at the end of the maze. As he predicted, the rats who had food waiting for them zipped through the maze faster and made fewer errors than the rats who were not rewarded for completing the maze. Curious whether the unrewarded rats had "learned less" than their rewarded counterparts, Tolman put some food at the end of the maze and sent the unrewarded group through. To his surprise, the unrewarded rats had not learned less; they just were not motivated to show him what they knew. Immediately after finding food at the center, the unrewarded rats started sprinting through the maze. The moral of the story here was an important one: An organism's outward behavior does not always tell the whole story. By figuring out ways to manipulate the animal's unobservable internal state, behaviorists were able to make inferences about animals' cognition and subjective experiences.

The results obtained by psychologists like Tolman and Watson could not have been obtained by naturalists because of the strict procedures they used. Similarly, the realness of the naturalists' observations could not have been obtained by the sterile, artificial environments created in the laboratory by psychologists. In this way, two very different sets of approaches offered valuable insights moving toward one unified field.

UNSUNG HEROES

Recognizing the significant role that context plays in the reception of any new discovery, it's no surprise that there are multiple proposed dates and founders. As mentioned earlier, the official beginning of animal cognition as a field is debatable. Though Romanes' publication of *Animal Intelligence* in 1882 marks one possible start of the field, we credit the formal beginning of the field to Margaret Floy Washburn, who published its first textbook, *The Animal Mind* (1908). Washburn organized decades of animal observations, published experiments, and methods into one textbook that was used to train multiple generations of researchers.

Science has historically been dominated by white men, and psychology at the turn of the 20th century, when Washburn was writing *The Animal Mind*, was no exception. However, through tireless efforts and, in some cases, pure luck, Washburn, a few of her female contemporaries, and also a few people of color were accepted into early psychology graduate programs. As two women researchers, one of whom is also a person of color, it's important to us as authors to pay tribute to the life and contributions of Margaret Floy Washburn and Charles Henry Turner, two pioneering individuals who are often left out of the history books.

Charles Henry Turner

After graduating valedictorian of his high school class, Charles Henry Turner (1867–1923) went on to college, where he published his first paper at just 24 years old (Turner, 1891). When a shorter version of that three-part paper appeared in *Science* (Turner, 1892a), Turner became one of, if not the first, African Americans to be published in this highly prestigious journal (Abramson, 2006). As another first, Turner was the first African American to earn a graduate degree from the University of Cincinnati in 1892. That same year he produced three other publications on the diverse topics of grapevine leaf growth (Turner, 1892b), aquatic crustaceans (Turner, 1892c), and variations in spiderweb formation (Turner, 1892d).

Not satisfied with a master's degree, he made history again studying zoology and became the first African American to earn a PhD from the University of Chicago. Even with a PhD from the school and dozens of publications, when Turner applied for a professorship at his alma mater, he was rejected. Whether his subsequent decision to teach at the high school level was forced or a preferred choice is unknown. Some (e.g., Abramson, 2009) speculate that despite earning the admiration and respect of white colleagues in the laboratory, institutional racism at the administrative level kept him from achieving his goal of a permanent professorship. Given Turner's tremendous publication record—more than double that of many of his white male colleagues' (Scarborough & Furumoto, 1987)—the fact he was passed up indicates race was very likely a significant factor.

Undeterred, Turner set up a productive research laboratory at Sumner High School, an all African American school. He lived, breathed, and worked animal cognition—going so far as to name one of his sons Darwin Romanes Turner. As a high school teacher, much of Turner's work focused on insects, including spatial navigation in ants and sensory systems in honey bees and moths. As Abramson (2006) points out, Turner's productivity was even more astonishing given the many barriers he faced as a high school teacher, including "few formal laboratory facilities, no easy access to research libraries, no opportunity to train research students at the undergraduate or graduate level, heavy teaching loads, low pay, and restricted research time" (p. 42). Facing these challenges, Turner persevered, contributing information about the behavior of dozens of different species to the scientific literature. In some cases, Turner pioneered methodologies and techniques that are still used—and sometimes misattributed to his more well-known white contemporaries (Abramson, 2003, 2006).

Though Turner used the laboratory-based method made popular by behaviorists at the time, his language and interpretations of behavioral findings reflected a researcher who was willing to make inferences about the minds of his subjects. For example, in his discussion of cockroaches running mazes, Turner (1913) wrote, "Some few roaches, after making several attempts to find an exit from the maze, stop trying and act as though they have given up all hope of succeeding" (p. 355). When he observed a cockroach rush to the edge of the maze and pause prior to jumping off, he concluded, "At times the roach acts as though experiencing the emotion psychologists call will" (p. 361). While Turner was willing to discuss animals' mental processes like "giving up" and voluntary decision making, he balanced it with the kind of laboratory objectivity that was championed by the behaviorist movement of his time.

Not betraying his naturalist roots, other lines of Turner's research were more observational in their methods. In one study, he removed the queen wasp from a small nest of 9 pupae and 15 larvae in order to learn how never seeing an adult in the colony affected subsequent generations' behavior (Turner, 1912). Among other discoveries, this 6-week-long detailed behavioral study allowed him to conclude that proper construction of a wasp nest may not be an entirely instinct-driven process.

In addition to contributing internationally renowned research during a time of blatant racism, Turner was also committed to the issue of education for African American youth. In one publication, for example, he defends the importance of teaching biology in African American schools as a way to "win the respect of other races" (Turner, 1897, p. 2). For his sustained civil rights work, at least four predominantly African American–serving institutions today bear his name. Sadly, Turner's prolific research program at Sumner High School lasted only 15 years, and he died shortly after retiring in 1922, but his name and legacy live on. The Animal Behavior Society has furthered Turner's advocacy work by offering the Charles H. Turner Award, a conference attendance scholarship for underrepresented college students interested in animal behavior, and Turner is the focus of the 1997 children's book *Bug Watching With Charles Henry Turner*, which introduces the next generation of scientists to animal behavior.

Margaret Floy Washburn

Like Turner, Margaret Floy Washburn (1871–1939) also set herself apart from her peers from an early age, first by starting college at the age of 16. During her senior year of college, Washburn turned her attention to pursuing psychology, where, also like Turner, she faced discrimination. Because Washburn was a woman, she could not attend her preferred graduate program at Columbia University. Instead, she was allowed only to audit courses and work in an experimental psychology laboratory (McHenry, 1980). Washburn was lucky in that she considered her research adviser, James McKeen Cattell, to be "a lifelong champion of freedom and equality of opportunity" (O'Connell & Russo, 1983, p. 17), but when even he could not get her into Columbia, she moved on to Cornell University, where she became the first woman *ever* to be awarded a PhD in psychology. Her human psychology research focused on the relationship between visual imagery, movement, and perception. Her publications included topics like distance judgments (Washburn, 1894), object recognition (Washburn, 1897), logic (Washburn, 1898), and preferences for color combinations (Washburn, MacDonald, & Van Alstyne, 1922), to name a few.

Eventually, however, Washburn grew dissatisfied with the extremes of structuralism (all mental) and behaviorism (all outward behavior) and decided to develop her own dualist theory regarding the connection between motor movement and social consciousness (i.e., awareness). Specifically, she hypothesized that development of social consciousness did not arise in social species from imitation learning, but rather that our physical social behaviors during early childhood actually allow social consciousness to develop (Washburn, 1903). As you might suspect, Washburn's controversial body-informing-mind theory would have made Descartes roll in his grave! We encourage the reader to consult Martin (1940) for a fantastic synthesis of Washburn's major contributions.

Washburn settled at Vassar College, where, with the assistance of senior psychology students working in her laboratory, she published more than 70 articles in the *American Journal of Psychology*. Due to her gender, Washburn's climb to the top of the ranks as a pioneering experimental psychologist was not a smooth one. Her strategy was to essentially ignore the established social norms of how women scientists, and women in general, were supposed to behave around men. Vassar College's president during Washburn's tenure recounted, "Miss Washburn had been intrepid enough to invade the sacred precinct of the men's smoker at psychological meetings. Marching uninvited into its midst, she had sat down and lighted a cigar. None questioned her privilege to enjoy the smoker thereafter" (MacCracken, 1950, p. 70). Naturally, only someone this brazen could take the reins and pull together years of research to write the first textbook for an entirely new field, and that's just what Washburn did.

The Animal Mind

While Washburn's active research into motor movement, sensory perception, and social consciousness was more than enough to cement her as what some consider to be the most accomplished female psychologist (Scarborough & Furumoto, 1987), it was her textbook *The Animal Mind* that catapulted her to fame in the animal cognition world. The motivation for writing came from her theory that consciousness was not unique to humans. Dissatisfied with simply theorizing about the topic, Washburn began conducting experimental investigations to test her hypothesis.

In 1908, she published *The Animal Mind*, a collection of experimental data on topics like sensory systems, learning, tool use, motivation, and subjective experience. Because Washburn believed cognition could be observed throughout the animal kingdom, she included dozens of diverse species, like amoebas, insects, dogs, shellfish, rats, cows, and salamanders, to name a few. The textbook also outlined appropriate methodology, experimental design and apparatuses, and the reminder of the challenge of inferring an individual's internal experience from their external behavior. Washburn noted in her book that in order to draw conclusions about much of science, inferences must be made, including in the understanding of subjective experiences of even humans. She also said that every species' body is different from the human body, so it is not fair to conclude an animal is not capable of something because their anatomy may not map onto our own. These sorts of thoughtful considerations, along with the presentation of empirical data for such a wide range of species, were very well received and marked the official start of the field. *The Animal Mind* trained generations of budding animal cognition researchers for three additional editions before Washburn's death in 1939.

HISTORICAL CHALLENGES IN THE FIELD

In the next chapter, you will read about common terms and specific methodologies used in the formal study of animals' cognitive abilities. Interestingly, despite a century's worth of research, discourse, and technological advancements, many of the conceptual challenges faced by animal cognition pioneers of the early 20th century are still very much issues that contemporary researchers face today. These concerns are the tendency to overextend human traits to animals and the undeniable human superiority bias that pervades Western society. Similar challenges plague animal cognition researchers at the level of interpreting their findings, but we save that for Chapter 2, Theoretical and Methodological Approaches to Animal Cognition.

Anthropomorphism

Imagine (or perhaps recall an actual memory if this has ever happened to you) that you return home from work and your dog does not greet you excitedly at the door. Instead, she peeks around the corner at you, her tail between her legs, head down in a most submissive posture, avoiding eye contact. Something is definitely going on, but you cannot determine what. You enter the living room to find it absolutely destroyed—couch cushion stuffing, houseplants, potting soil, and shredded newspaper litter the floor. Now, how might you interpret your dog's behavior? If you're thinking *guilt, remorse, embarrassment,* or *shame,* you would not be alone. Do a quick Internet search for *dog shaming memes* if you do not believe us. Attributing human traits to a nonhuman entity, termed *anthropomorphism,* is rampant in how most people think about and interact with animals, whether they are pets or not (Horowitz & Bekoff, 2007). Some researchers even speculate that our tendency to see commonality between human and animal behavior could be the result of thousands of years of human evolution. Specifically, it may have been an advantage to our ancestors to better be able to (or perhaps think they were better able to) predict an animals' behavior (e.g., Mithen, 1996).

The word *anthropomorphism* existed for a long time, mainly to describe gods and celestial beings as humanlike, but its likely first usage for animals was made in 1858. This was just around the time Darwin began promoting his ideas that would be published the following year in *The Origin of Species*. But it was not Darwin himself who used the term. Philosopher and physiologist George Henry Lewes used it in his cautionary warning against doing it. As we saw earlier, neither Darwin nor some of his followers like Romanes saw much of a problem with anthropomorphizing. They did so liberally and with their own justification. It is worth noting that even Romanes had a limit, admitting that in drawing inferences about insect behavior, there is a "progressive weakening" of the ability to find similarities between human and nonhuman behavior (p. 9).

When behaviorism cornered the market on psychology at the turn of the 20th century, Watson and others outright rejected the liberties early animal researchers had taken in extending Darwin's continuity-of-species assumption to animals' minds. In many ways, their strong rejection makes sense given Watson's goal was to make psychology an objective science, which neither studied nor drew conclusions about anyone's mental processes, human or nonhuman. While some at the time rebelled against Watson's strict behaviorism, it was not until the 1950s or so that researchers "brought the mind back into experimental psychology" (Miller, 2003, p. 141). The trickle-down effect from human psychology to the animal cognition world was slow, but by the 1970s, studying animals' cognitive abilities had regained traction and approval (Wynne, 2007). If you look carefully in this and subsequent chapters, you'll find a gap in published animal cognition work from around 1940 to 1970. This reflects behaviorism's stifling effect on the field.

Whether or not anthropomorphism has a place in animal cognition research depends on who you talk to. Some argue that anthropomorphism represents the "prescientific," subjective way of thinking that experimental psychologists have tried to avoid engaging in (e.g., Wynne, 2007). Wynne makes a good point when he says that since Darwin's time, we have uncovered many lawful and predictable relationships about animal behavior that do not require us to anthropomorphize anymore. For example, in the case of the hypothetical dog destroying my house, I could stop and remember back to 1 month ago when my dog tipped over a potted plant and I stood over him and loudly scolded him. Now, recalling that memory, the behavior I saw at the front door could be explained theoretically as a normal fear response, with my dog anticipating punishment for tipping over the potted plant. My interpretation no longer needs to connect how I would feel in that situation and arrive at guilt. Instead, we can use what we have learned about predictable relationships between the environment, behavior, and outcomes, in order to explain what I saw.

Another reason why interpretations grounded in anthropomorphism have been challenged is because they offer a convenient way to stop searching for answers by applying an easy label to behaviors that look familiar (e.g., Blumberg & Wasserman, 1995). According to this argument, once I identify a behavior that looks familiar enough for me to interpret it, such as how *I* would feel and behave if *I* got caught doing something wrong, I can stop investigating other possible explanations. Who knows, maybe my dog was behaving that way because she was actually sick from eating the plant she tipped over, but I was too busy posting photos of her on social media to notice. A label is just a label, not necessarily an actual explanation, but by anthropomorphizing my dog's behavior, I can stop at the label of *guilt* and snap a funny photo.

But wait, what if my dog *was* experiencing guilt and I dismissed it because I channeled my inner John Watson and ignored it? Famed ethologist Konrad Lorenz (1991) said, "If we feel ourselves emotionally affected by the behavior of an animal, it is a clear indication that we have intuitively discovered a similarity between its behavior and human behavior. We should not conceal this in our description" (pp. 260–261). While we advocate for considering behavior theory interpretations first, we also neither ignore nor avoid investigating cognitive similarities between humans and nonhuman animals. If we did, our interpretations of animal behavior could unnecessarily over- or underestimate their actual abilities, just because we are afraid of drawing comparisons to human behavior.

The second half of the 20th century produced two other scientists whose views on anthropomorphism are also important to consider. Donald Griffin (1976) was one of the first who dared to suggest that scientists might actually benefit from an anthropomorphic lens of interpretation (Wynne, 2005). For example, given the experimental research that had already occurred, Griffin thought it was no longer accurate to insist that animals had no awareness of the world at all, and if that was the case, anthropomorphism with regard to awareness would be beneficial to future research. A decade later, Gordon Burghardt coined the term *critical anthropomorphism* for the use of data from multiple sources—including anthropomorphic ones—but which could then be formed into appropriately testable hypotheses (Burghardt, 1985). Research on reconciliation in chimpanzees (*Pan troglodytes*; de Waal, 1999) and self-recognition in a variety of species (Gallup, 1970), for example, are both areas of study in which discoveries might not have been possible without at least some degree of anthropomorphism (Burghardt, 1985).

Human Uniqueness

Underlying anthropomorphism is a fundamental belief in commonality, the assumption that there is something similar between humans and other animals. While it's normal to anthropomorphize, it must be applied carefully because we cannot ask the dog, "Are you cowering because you feel ashamed about destroying my living room?" The argument by analogy forces animal cognition researchers to think more deeply about animal behavior.

See if you can identify the problem with the following:

(1) Turkeys and humans are both animals.
(2) Turkeys are food.
Therefore, humans are food.

In the first two statements, claims are made. Turkeys and humans are, indeed, both animals, and turkeys are considered food in the United States. The analogy breaks down at the conclusion. Just because turkeys and humans share some similar features, it does not mean that we can confidently extend those similarities to assume similarities across all possible areas.

Here is another set of statements that are more relevant to our discussion:

(1) Humans and chimpanzees are closely related.
(2) Humans can think abstractly.
Therefore, chimpanzees can think abstractly.

Darwin's assumption of continuity of species was invaluable to science, but it requires a strong need for checks and balances in logic and conclusions. This means every good animal cognition researcher must toe the line of having an open mind and being a healthy skeptic. Inappropriate arguments by analogy can muddy the waters of interpretation when we observe animals' outward behavior. It's important to remember that just because an animal exhibits some outward behavior or possesses some brain structure that, in humans, corresponds to a particular mental state or cognitive ability, it does not mean we can jump to conclusions without serious consideration of alternative explanations. Further, each species has been acted upon by thousands, if not millions, of years of evolution, meaning assumptions should not be made about even the most familiar of behaviors. Try smiling at a baboon (*Papio* spp.) sometime, and when your friendly gesture is met with aggression (because, actually, smiling is a threat gesture in almost every nonhuman species), you'll understand what we mean about making assumptions.

You're likely reading this book because you're genuinely interested in learning more about the cognitive abilities of other species. While *you* may be open minded to the possibility that animals can manufacture and use tools, reflect on their own knowledge, or have culture, many people are not. For some, the threat of not being "special" anymore—whatever that means!—is a difficult pill to swallow. By engaging in what primatologist Frans de Waal calls anthropodenial, some people completely shut out the possibility that animals are capable of demonstrating evidence of complex social and cognitive abilities. Anthropodenialist perspectives keep humans at the top of an imaginary hierarchy. This hierarchy justifies our treatment of animals in ways that we would never treat humans, deepening the line in the sand that Descartes and others drew hundreds of years ago. If we found out other species are more similar to us than we'd like to think, it might force serious changes in our society, which many people simply are not ready or willing to think about.

In considering the roots of the line in the sand, the "humans are special" bias, we must go back further than Descartes to explore a system that is foundational to many societies: religion. Specifically in Westerns countries, Christianity has played an undeniable role in why humans tend to create a species hierarchy with humans at the top and why, for many years, the common belief was that humans were fundamentally different from other species (Singer, 2002). Christianity certainly impacted Descartes' views and was a source of fear for Darwin as he prepared to publish *On the Origin of Species*. Thus, even with scientific rigor imposed by methodology (Chapter 2) and careful interpretations of findings, animal cognition researchers must also be aware of deep-seated biases at the societal level that may lead others to reject their findings.

Things have gotten much better in the past 50 years or so. Many of the methodological challenges are still present, but the field has grown and refined itself into a reputable science. Researchers like Jane Goodall and Irene Pepperberg have captivated audiences with their investigations, and popular science books on animal cognition have become *New York Times* best sellers. Keeping in mind the influence that society and its values has on science, the field's second century should be an exciting one!

ANIMAL SPOTLIGHT: KANZI

We humans are fascinated by our primate relatives. But why? Of all animal species, humans rate apes and monkeys as having life experiences and cognitive abilities that are most similar to those of humans (Eddy, Gallup, & Povinelli, 1993). Since we're on the topic, the terms *monkey* and *ape* are often used synonymously, but they refer to two groups that have about 40 million years' worth of differences. Generally speaking, compared to apes, monkeys are smaller, have visible tails, and conical rather than barrel chests. Dogs also have conical chests, meaning deeper than they are wide. Humans have barrel chests, wider than they are deep, so our shoulder blades allow us to perform our morning stretch with our arms extended out to the side. Other apes can do this, too. And there you have it, a bit of Primates 101!

Pioneering work in the mid-20th century focused on teaching language to apes in a variety of settings (i.e., laboratory and home) using English speech, American Sign Language, and made-up languages. It was not until the 1980s that a promising new method arose, one that took inspiration from Japanese primatologists (Segerdahl, Fields, & Savage-Rumbaugh, 2005). By keeping at the forefront what makes a human *human*, lead researcher Sue Savage-Rumbaugh stunned the world with her decades-long investigations of one very special ape named Kanzi.

Born into captivity in 1980, Kanzi the bonobo (*Pan paniscus*)—a subspecies of chimpanzee—was brought to Georgia State University's Language Research Center (LRC) in Atlanta at 6 months old. At the LRC, handlers teach a made-up language called Yerkish, which consists of hundreds of abstract symbols (i.e., lexigrams) that have meaning, just like human languages have arbitrary symbols that stand in the place of real objects, actions, and events in the world. For more than 2 years, Kanzi accompanied his adopted bonobo mother, Matata, to her Yerkish sessions, and much to the surprise of the researchers, Kanzi spontaneously learned to communicate with the lexigrams purely by observation (Savage-Rumbaugh, Shanker, & Taylor, 1998).

Savage-Rumbaugh recognized that like a human child learning language, Kanzi had learned Yerkish naturally by observing and interacting. Whereas all prior attempts to teach language to apes had used very rigorous, structured "language-training" sessions and they had all failed, Savage-Rumbaugh and her colleagues embarked on a different path: "If we talk with apes as we talk with children—taking for granted that understanding will appear—then the apes will begin to understand us and even speak to us" (Segerdahl et al., 2005, p. 197). Feeling that language is an important part of culture, Savage-Rumbaugh combined the scientific names for bonobos and humans to develop what her team referred to as *Pan/Homo* culture, where bonobos and humans interact and influence each other, while retaining important features of their own species. Under this new framework, Kanzi became the focus of a variety of different types of studies, all of which have taught us about the evolution of language, tool making, culture, and other behaviors typically thought to be uniquely human.

Kanzi is most famous both for his impressive comprehension of English and for his sophisticated use of Yerkish lexigrams (i.e., arbitrary visual symbols) to communicate. Once he showed promise

CHAPTER 1 HISTORICAL PERSPECTIVE ON ANIMAL COGNITION

as a Yerkish user, Savage-Rumbaugh began to wonder if he might help her understand the language learning process. To mimic how children learn language, Kanzi and his handlers worked with the lexigrams in natural contexts like walking through the woods, rather than in the laboratory. The result of this more natural environment was that Kanzi's Yerkish skills flourished, with a working vocabulary of over 250 lexigrams referencing different nouns, verbs, and adjectives, including one of our favorites, BURRITO.

To illustrate Kanzi's sophisticated use of lexigrams, we summarize a scene from the 1993 documentary *Kanzi: An Ape of Genius*. To set the scene, Kanzi has been separated from the rest of his group and, in particular, his adopted mother, Matata.

> Kanzi approaches Savage-Rumbaugh who asks him "Did you want something?" Kanzi walks to the door and points to the keyhole, followed by pointing to the KEY lexigram. He gestures to Savage-Rumbaugh, then points out GROUP ROOM, KEY, KEY, MATATA, and GOOD. When Savage-Rumbaugh verbally interprets his lexigram request back to him, Kanzi vocalizes positively.

Some critics claim that by not working with Kanzi in a more structured way, situations like the preceding one are biased and do not offer reliable evidence of Kanzi's linguistic abilities (e.g., Kulick, 2017). Nonetheless, Savage-Rumbaugh defended her choice to sacrifice scientific rigor and objectivity in order to immerse Kanzi in a realistic language environment. Further, because of Kanzi's unconventional learning environment, Savage-Rumbaugh strongly believed that his communication abilities reflected true language. Coupled with her unconventional methods, this blow to one of the pillars of human uniqueness earned Savage-Rumbaugh a great deal of criticism (Savage-Rumbaugh & Lewin, 1994).

Though they did not like the idea of "testing" Kanzi's language abilities just to appease critics, Savage-Rumbaugh and her team knew they needed to offer more objective evidence. In order to show Kanzi's comprehension of more than 1,000 English words, Savage-Rumbaugh conducted some rigorous experiments. In them, she wore a mask to prevent her eyes and body gestures from giving Kanzi clues and found that he could perform commands he had never heard before, like, "Can you put your shirt in the refrigerator?" The fact that he could perform nonsense commands showed that English words had meaning to him (Segerdahl et al., 2005). Compared to a

2-year-old girl nicknamed Alia, Kanzi (8 years old at the time) correctly responded to 72% of nonsense commands like the preceding one, compared to Alia's 66% (Savage-Rumbaugh et al., 1993). Rather than framing their results as Kanzi responding to human language "better" than a 2-year-old, Savage-Rumbaugh and colleagues framed their results around the similarities of both individuals' performance. Specifically, they noted how impressively both performed given the required complexity of processing. For example, both Alia and Kanzi responded correctly to "Give the knife to [person]" and "Can you knife the sweet potatoes?" even though the word *knife* could be used both as a noun and verb. In another surprisingly impressive account, Kanzi demonstrated innovation when he responded to "Put the water on the carrot" by tossing a carrot outside into the rain (Savage-Rumbaugh et al., 1993).

Along with his contribution to our understanding of language, Kanzi has also taught us about the manufacturing and use of stone tools by our human ancestors. Stone flaking involves fracturing rocks with a hammerstone with the intention of creating sharpened pieces (i.e., flakes), which are used for a variety of purposes (Toth, Schick, Savage-Rumbaugh, Sevcik, & Rumbaugh, 1993). While plenty of animal species manufacture tools, many argue humans are the only species that craft stone tools. Paleoanthropologists researching the cognition and motor skills necessary for stone tool making observed Kanzi over the course of many years to see how his performance would stack up to modern humans and, presumably, humans' ape-like ancestors (Schick, Toth, & Garufi, 1999).

In one study, Kanzi was shown the basics of how to flake stones by a human, who then used a flake to cut a cord on a box and retrieve a food reward inside. After some demonstrations, he was given rocks, the fastened box, and a bit of praise for his efforts. Through trial and error, Kanzi learned to create stone flakes via the method he had observed. He also invented his own method for creating stone flakes by throwing large rocks onto the floor or other stones to fracture them. Over time, Kanzi produced hundreds of different stone flakes, successfully used them to open the box on dozens of occasions, and appeared to develop his own unique way of flaking stones to get around the physical challenge of controlling the hammerstone. On one hand, while Kanzi made "significant and rather startling progress" (Toth et al., 1993, p. 7), on the other, his skill level was far behind that of our human ancestors in the Stone Age. For example, he never spontaneously tried to modify the flakes he made, such as by sharpening them, though this could be because of differences in manual dexterity between humans and bonobos. Taken together, these findings tell paleoanthropologists that by the Stone Age, our ape-like human ancestors had likely already experienced evolutionary pressures that made them cognitively, anatomically, and behaviorally different when it came to skills like stone tool making.

After 25 years of contributing to animal cognition research at the LRC, Kanzi moved to what is now the Ape Cognition and Conservation Initiative (ACCI) in Des Moines, Iowa, and he "retired" from the vast majority of research by 2013. By giving Kanzi a part human–part bonobo life, Savage-Rumbaugh and her colleagues believe he was the first ape to learn to communicate with humans without active training (Cerrone, 2018). The result was an ape who could report on past events (e.g., "MATATA BITE" when asked about a wound on his hand), create novel utterances (e.g., "BAD SURPRISE" when he yanked a pillow out from under a sleeping person), and respond appropriately to complex commands (e.g., "You can have some cereal if you give Austin your monster mask to play with"; Savage-Rumbaugh et al., 1998). Kanzi's story illustrates the importance of many of the topics covered in this chapter, including anthropomorphism, being a healthy skeptic with an open mind, and checking our human superiority biases at the door when it comes to evaluating animals' cognitive abilities.

HUMAN APPLICATION: ANIMAL MODELS

When we think of Charles Darwin's most impactful contribution, the obvious comes to mind—his theory of evolution by natural selection. For most people, Darwin's theory of evolution might seem irrelevant to their everyday lives. However, Darwin's conclusions gave a definitive green light to studying the animal body as a way to better understand and treat the human body. Whether we agree with the practice of using animals in biomedical research or not, the truth is that many people around the world have been positively impacted by animal models.

Animal models have been around since at least the 6th century BCE (Ericsson, Crim, & Franklin, 2013), including Aristotle using chicken embryos to understand human fetal development and medieval physicians perfecting surgical techniques on dogs before using them on humans. The practice really took off by the late 19th century, likely due in part to Darwin. His ideas about emotional, behavioral, and cognitive similarities between humans and animals justified animals being used as stand-ins for humans in modern medicine and psychiatry. Animal models have offered insights into everything from human food allergies (Helm, 2006), to Alzheimer's disease (Webster, Bachstetter, Nelson, Schmitt, & Van Eldik, 2014), to even treating severe mental disorders like schizophrenia (Hamm, Peterka, Gogos, & Yuste, 2017). Believe it or not, early medications for depression were actually developed using guinea pigs (*Cavia porcellus*; Ericsson et al., 2013).

When we think of animal models, we usually envision primates and rodents. Because of this, we chose to share some lesser known animal model species with the hopes that you will gain a greater appreciation for (1) how surprising and controversial the idea of an assumption of continuity of species might have been at the time of Darwin's publications and (2) just how similar humans are to many other species. Of course, as we have seen, caution must be taken in drawing this latter conclusion, and so we have also included a few challenges to using animal models along the way.

Medical Models

Axolotl (*Ambystoma mexicanum*)

Unlike most animals that heal by scarring over a wound, this salamander has the amazing superpower of regenerating entire limbs, organs, and even its spinal cord following damage or amputation (Roy & Gatien, 2008). So far, researchers think this occurs because the axolotl's bone, skin, and other tissue cells revert to stem cells when they're damaged. A stem cell is a generic cell that can develop into bone, skin, or other tissue cell as needed (Erickson & Echeverri, 2018). Following injury, the axolotl's correct genes are "turned on," and its stem cells create a new leg, tail, heart, and so forth.

Axolotls and humans have not shared a common ancestor for 350 million years, but medical scientists see the value in learning more about their regenerative abilities. With increasing rates of diabetes, cancer, and other shifting health demographics, the number of amputees in America is predicted to double from 1.6 million in 2005 to 3.6 million by 2050 (Ziegler-Graham, MacKenzie, Ephraim,

Travison, & Brookmeyer, 2008). If we can figure out how the axolotl's body regenerates, we might be able to apply that understanding to help millions of people. Even if regeneration is not possible, harnessing the axolotl's scar-free healing would also substantially improve body image and quality of life for humans following major surgery or injury.

Zebrafish (*Danio rerio*)

In a survey investigating how similar humans felt they were to a list of different animals, fish ranked barely above the "Not Similar at All" cutoff (Eddy et al., 1993). The medical community knows the exact opposite. Zebrafish, a type of minnow, have been used since the mid-1990s to model a variety of human diseases and physiological functions, including embryo development, color vision, muscular dystrophy, and immune system functioning (Ericsson et al., 2013). Zebrafish are also used as a cancer model.[1] Oncologists can isolate specific genes in the fish and screen them for cancer, a technique that can tell us more about how cancer grows, how environmental factors can slow or worsen cancer spread (Yen, White, & Stemple, 2014), and, perhaps one day, how to stop "cancer genes" from being expressed (Coel et al., 2011).

Brain Models

Fruit Fly (*Drosophila melanogaster*)

Parkinson's disease (PD) disrupts brain functioning and is generally observed in older adults. People with PD have problems with motor movement and tremors, but memory loss, cognitive impairment, and mood disorders are also common. While there is no cure for PD, the motor impairment associated with PD can be treated by giving patients a drug called L-DOPA, which replenishes PD patients' severely low dopamine levels. You may have heard of dopamine as being a chemical your brain produces that is relevant to pleasure, happiness, and reward. Interestingly, dopamine is also involved with

---

1. Although more recently, the fashionable model is the naked mole rat (*Heterocephalus glaber*) due to their unusually high cancer resistance (Taylor, Milone, & Rodriguez, 2016). Only six cases of spontaneous cancer (all in animals under human care) have ever been documented.

smoothing out motor movements as well as learning and memory. Enter the fruit fly.

In 2000, Feany and Bender genetically modified fruit flies to produce a mutated human protein that is found in PD patients. The result was flies that had motor movement problems that were similar to what human PD patients experience. Further, by increasing dopamine levels in PD fruit flies, their motor movement improved, just like in humans (Valadas, Vos, & Verstreken, 2015). The fact that PD symptoms and treatments seem to affect humans and fruit flies similarly suggests that scientists can test PD medications on the flies and reasonably expect them to have similar effects on humans.

Pond Snail (*Lymnaea stagnalis*)

Turn over a rock at the beach to find one of psychology's models for learning and memory, the pond snail. While the human brain might have an impressive 100 billion neurons, the pond snail's 11,000 can tell us more about ourselves than we may think. Believe it or not, snails and slugs are capable of remembering and responding in predictable ways to stimuli they have experienced before. For example, with repeated trials, snails can learn to avoid a particular food if it has been paired with a negative outcome, the same way a person might avoid oysters forever if they get sick from a bad batch. However, by manipulating chemicals in the snails' brains before this taste aversion training occurs, researchers can stop their brains from forming a memory to avoid the food (Murakami et al., 2013). Being able to chemically disrupt the learning and memory process might sound like the makings of a sci-fi film, but it has serious real-world implication. What if we were able to manipulate or even erase memories at will? Who would be allowed to do this? What legal or ethical implications would need to be considered?

Evaluating Animal Models

Animal models will not be going away anytime soon. However, there are many challenges to using them. Practically speaking, human brains function with greater complexity than those of other species (including other primates), which may make it difficult to accurately model human psychological processes in animals. Further, especially in the case of developing treatments for neuropsychological disorders like PD, medications tested in animals may reduce outward symptoms such as tremors, while nothing is known about the drug's effects on the subjective experience of the animal (Nestler & Hyman, 2010). It would not be until the drug was tested on an actual human that the mental effects of the drug would be known. This poses understandable ethical concerns.

On the other side of the coin, ethical considerations for the animals involved with such research also pose challenges. Some argue that since we are not 100% sure diseases such as depression or schizophrenia affect humans and animals similarly, it's unethical to subject the animals to these diseases in the first place. These same issues arise in cancer research, where zebrafish appear to model some, but not all, forms of cancer (Yen et al., 2014). Without understanding why, is it acceptable to subject the animals to experimentation? As one medical professional put it:

> The debate on the ethics of animal research has caught the researchers in a logical trap: in order to defend the usefulness of research they must emphasize the similarities between the animals and the humans, but in order to defend it ethically, they must emphasize the differences. (Flossos, 2005, p. 4)

While Darwin's continuity of species gave researchers firm ground to walk on when studying human processes in animals, we must always consider the preceding quote as we use animals to pursue our own medical and psychological advancements.

REFERENCES

Abramson, C. I. (2003). *Selected papers and biography of Charles Henry Turner, 1867-1923: Pioneer in the comparative animal behavior movement.* Lewiston, NY: Edwin Mellen Press.

Abramson, C. I. (2006). Charles Henry Turner: Pioneer of comparative psychology. In D. A. Dewsbury, L. T. Benjamin, & M. Wertheimer (Eds.), *Portraits of pioneers in psychology* (pp. 37–49). New York, NY: Psychology Press.

Abramson, C. I. (2009). A study in inspiration: Charles Henry Turner (1867–1923) and the investigation of insect behavior. *Annual Review of Entomology, 54*, 343–359. doi:10.1146/annurev.ento.54.110807.090502

Blumberg, M. S., & Wasserman, E. A. (1995). Animal mind and the argument from design. *American Psychologist, 50*, 133–144. doi:10.1037/0003-066X.50.3.133

Boakes, R. A. (1984). *From Darwin to behaviorism: Psychology and the minds of animals.* Cambridge, UK: Cambridge University Press.

Burghardt, G. M. (1985). Animal awareness. Current perceptions and historical perspective. *The American Psychologist, 40*(8), 905–919. Retrieved from http://www.ncbi.nlm.nih.gov/pubmed/3898938

Cerrone, M. (2018). Umwelt and ape language experiments: On the role of iconicity in the human-ape pidgin language. *Biosemiotics, 11*(1), 1–23. doi:10.1007/s12304-018-9312-4

Coel, C. J., Houvras, Y., Jane-Valbuena, J., Bilodeau, S., Orlando, D. A., Battisti, V., … Zon, L. I. (2011). The histone methyltransferase SETDB1 is recurrently amplified in melanoma and accelerates its onset. *Nature, 471,* 513–517. doi:10.1038/nature09806

Darwin, C. R. (1871). *The descent of man and selection in relation to sex.* London, UK: John Murray.

Darwin, C. R. (1872). *The expression of the emotions in man and animals* (1st ed.). London, UK: John Murray.

de Waal, F. B. M. (1999). Anthropomorphism and anthropodenial: Consistency in our thinking about humans and other animals. *Philosophical Topics, 27,* 255–280. doi:10.5840/philtopics199927122

Ebbinghaus, H. (1908). *Psychology: An elementary textbook.* Boston, MA: Heath.

Eddy, T. J., Gallup, G. G., & Povinelli, D. J. (1993). Attribution of cognitive states to animals: Anthropomorphism in comparative perspective. *Journal of Social Issues, 49,* 87–101. doi:10.1111/j.1540-4560.1993.tb00910.x

Erickson, J. R., & Echeverri, K. (2018). Learning from regeneration research organisms: The circuitous road to scar free wound healing. *Developmental Biology, 433,* 144–154. doi:10.1016/j.ydbio.2017.09.025

Ericsson, A. C., Crim, M. J., & Franklin, C. L. (2013). A brief history of animal modeling. *Missouri Medicine, 110,* 201–205.

Feany, M. B., & Bender, W. W. (2000). A *Drosophila* model of Parkinson's disease. *Nature, 404,* 394–398. doi:10.1038/35006074

Flossos, A. (2005). Ethical issues in animal research. *The Greek Journal of Perioperative Medicine, 3,* 1–5.

Gallup Jr., G. G. (1970). Chimpanzees: Self-recognition. *Science, 167,* 86–87. doi:10.1126/science.167.3914.86

Gould, S. J. (2002). *The structure of evolutionary theory.* Cambridge, MA: Belknap Press.

Griffin, D. R. (1976). *The question of animal awareness: Evolutionary continuity of mental experience.* New York, NY: Rockefeller University Press.

Hamm, J. P., Peterka, D. S., Gogos, J. A., & Yuste, R. (2017). Altered cortical ensembles in mouse models of schizophrenia. *Neuron, 94,* 153–167.e8. doi:10.1016/j.neuron.2017.03.019

Helm, R. M. (2006). Food allergy animal models. *Annals of the New York Academy of Sciences, 964,* 139–150. doi:10.1111/j.1749-6632.2002.tb04139.x

Horowitz, A. C., & Bekoff, M. (2007). Naturalizing anthropomorphism: Behavioral prompts to our humanizing of animals. *Anthrozoös, 20,* 23–35. doi:10.2752/089279307780216650

Kulick, D. (2017). Human-animal communication. *Annual Review of Anthropology, 46*, 357–378. doi:10.1146/annurev-anthro-102116-041723

Logue, A. W. (1978). Behaviorist John B. Watson and the continuity of the species. *Behaviorism, 6*(1), 71–79.

Lorenz, K. (1991). *Here I am—Where are you?* New York, NY: Harcourt Brace Jovanovich.

MacCracken, H. N. (1950). *The hickory limb*. New York, NY: Scribner's.

Martin, M. F. (1940). The psychological contributions of Margaret Floy Washburn. *The American Journal of Psychology, 53*, 7–18.

McHenry, R. (1980). *Famous American women: A bibliographical dictionary from colonial times to the present*. New York, NY: Dover Publications, Inc.

Miller, G. A. (2003). The cognitive revolution: A historical perspective. *Trends in Cognitive Science, 7*, 141–144. doi:10.1016/S1364-6613(03)00029-9

Mithen, S. (1996). *The prehistory of the mind: The cognitive origins of art, religion, and science*. London, UK: Thames and Hudson Ltd.

Murakami, J., Okada, R., Sadamoto, H., Kobayashi, S., Mita, K., Sakamoto, Y, … Ito, E. (2013). Involvement of insulin-like peptide in long-term synaptic plasticity and long-term memory of the pond snail *Lymnaea stagnalis*. *Journal of Neuroscience, 33*, 371–383. doi:10.1523/JNEUROSCI.0679-12.2013

Nestler, E. J., & Hyman, S. E. (2010). Animal models of neuropsychiatric disorders. *Nature Neuroscience, 13*, 1161–1169. doi:10.1038/nn.2647

O'Connell, A., & Russo, N. (1983). *Models of achievement: Reflections of eminent women in psychology*. New York, NY: Columbia University Press.

Power, D. (1898). *William Harvey*. London, UK: Longmans, Green & Company.

Romanes, G. J. (1882). *Animal intelligence*. London, UK: K. Paul, Trench.

Ross, M. E. (1997). *Bug Watching with Charles Henry Turner*. Minneapolis, MN: Carolrhoda Books.

Roy, S., & Gatien, S. (2008). Regeneration in axolotls: A model to aim for! *Experimental Gerontology, 43*, 968–973. doi:10.1016/j.exger.2008.09.003

Savage-Rumbaugh, E. S., Murphy, J., Seveik, R. A., Brakke, K. E., Williams, S. L., Rumbaugh, D. M., & Bates, E. (1993). Language comprehension in ape and child. *Monographs of the Society for Research in Child Development, 58*, 1–256. doi:10.2307/1166068

Savage-Rumbaugh, S., Shanker, S. G., & Taylor, T. J. (1998). *Apes, language, and the human mind*. New York: Oxford University Press.

Savage-Rumbaugh, S. S., & Lewin, R. (1994). *Kanzi: The ape at the brink of the human mind*. New York, NY: John Wiley & Sons.

Scarborough, E., & Furumoto, L. (1987). *Untold lives: The first generation of American woman psychologists*. New York, NY: Columbia University Press.

Schick, K. D., Toth, N., & Garufi, G. (1999). Continuing investigations into the stone tool-making and tool-using capabilities of a bonobo (*Pan paniscus*). *Journal of Archaeological Science, 26*, 821–832. doi:10.1006/jasc.1998.0350

Segerdahl, P., Fields, W., & Savage-Rumbaugh, S. (2005). *Kanzi's primal language: The cultural initiation of primates into language.* New York, NY: Palgrave Macmillan.

Singer, P. (2002). *Animal liberation* (3rd ed.). New York, NY: HarperCollins.

Taylor, K. R., Milone, N. A., & Rodriguez, C. E. (2016). Four cases of spontaneous neoplasia in the naked mole-rat (*Heterocephalus glaber*), a putative cancer-resistant species. *The Journals of Gerontology: Series A, 72*, 38–43. doi:10.1093/gerona/glw047

Tolman, E. C. (1948). Cognitive maps in rats and men. *Psychological Review, 55*, 189–208. doi:10.1037/h0061626

Toth, N., Schick, K. D., Savage-Rumbaugh, E. S., Sevcik, R. A., & Rumbaugh, D. M. (1993). Pan the tool-Maker: Investigations into the stone tool-making and tool-using capabilities of a bonobo (*Pan paniscus*). *Journal of Archaeological Science, 20*, 81–91. doi:10.1006/jasc.1993.1006

Turner, C. H. (1891). Morphology of the avian brain. *Journal of Comparative Neurology, 1*, 39–93, 107–133, 265–286. doi:10.1002/cne.910010307

Turner, C. H. (1892a). A few characteristics of the avian brain. *Science, 19*, 16–17. doi:10.1126/science.ns-19.466.16

Turner, C. H. (1892b). A grape vine produces two sets of leaves during the same season. *Science, 20*, 39. doi:10.1126/science.ns-20.493.39

Turner, C. H. (1892c). Notes upon Cladocera, Copepoda, Ostracoda and Rotifera of Cincinnati with description of new species. *Bulletin of the Scientific Laboratories of Denison University, 6*, 57–74.

Turner, C. H. (1892d). Psychological notes upon the gallery spider: Illustrations of intelligent variations in the construction of the web. *Journal of Comparative Neurology, 2*, 95–110.

Turner, C. H. (1897). Reason for teaching biology in Negro schools. *Southwestern Christian Advocate, 32*, 2.

Turner, C. H. (1912). An orphan colony of *Polistes pallipes* Lepel. *Psyche, 19*, 184–190. doi:10.1155/1912/67292

Turner, C. H. (1913). Behavior of the common roach (*Periplaneta orientalis* L.) on an open maze. *Biological Bulletin, 25*, 348–365. doi:10.2307/1536129

Valadas, J. S., Vos, M., & Verstreken, P. (2015). Therapeutic strategies in Parkinson's disease: What we have learned from animal models. *Annals of the New York Academy of Sciences, 1338*, 16–37. doi:10.1111/nyas.12577

Voltaire. (1901). *The works of Voltaire. A contemporary version, in 21 vols.* (Original work published 1764). Retrieved from https://oll.libertyfund.org

Washburn, M. F. (1894). The perception of distance in an inverted landscape. *Mind, 3*, 438–440.

Washburn, M. F. (1897). The process of recognition. *Philosophical Review, 6*, 267–274.

Washburn, M. F. (1898). The psychology of deductive logic. *Mind, 7*, 523–530.

Washburn, M. F. (1903). The genetic function of movement and organic sensations for social consciousness. *The American Journal of Psychology, 14*, 337–342. doi:10.2307/1412306

Washburn, M. F. (1908). *The animal mind; a text-book of comparative psychology.* New York, NY: The Macmillan Company.

Washburn, M. F., MacDonald, M. T., & Van Alstyne, D. (1922). Voluntarily controlled likes and dislikes of color combinations. *The American Journal of Psychology, 33*, 426–428. doi:10.2307/1413530

Watson, J. B. (1913). Psychology as the behaviorist views it. *Psychological Review, 20*, 158–177. doi:10.1037/h0074428

Webster, S. J., Bachstetter, A. D., Nelson, P. T., Schmitt, F. A., & Van Eldik, J. L. (2014). Using mice to model Alzheimer's dementia: An overview of the clinical disease and the preclinical behavioral changes in 10 mouse models. *Frontiers in Genetics, 5*, 88. doi:10.3389/fgene.2014.00088

Wynne, C. D. L. (2005). The emperor's new anthropomorphism. *The Behavior Analyst Today, 6*, 151–154. doi:10.1037/h0100066

Wynne, C. D. L. (2007). What are animals? Why anthropomorphism is still not a scientific approach to behavior. *Comparative Cognition & Behavior Reviews, 2*, 125–135. doi:10.3819/ccbr.2008.20008

Yen, J., White, R. M., & Stemple, D. L. (2014). Zebrafish models of cancer: Progress and future challenges. *Current Opinion in Genetics and Development, 24*, 38–45. doi:10.1016/j.gde.2013.11.003

Ziegler-Graham, K., MacKenzie, E. J., Ephraim, P. L., Travison, T. G., & Brookmeyer, R. (2008). Estimating the prevalence of limb loss in the United States: 2005 to 2050. *Archives of Physical Medicine and Rehabilitation, 89*, 422–429. doi:10.1016/j.apmr.2007.11.005

Theoretical and Methodological Approaches to Animal Cognition

hile animal cognition researchers may look strange playing parrot vocalizations on loudspeakers in the jungle or presenting gorillas with trays of colored shapes, there very much is a method to our madness. That method is the scientific method whose steps are to develop a research question, design appropriate methodologies, collect and analyze data, then share the findings with the scientific community. In this chapter, we present some considerations and methods for studying animal cognition with the hope that you, the reader, will use some of them in your future observations of animal behavior—both human and nonhuman!

TERMS AND CONSIDERATIONS

Animal cognition is a branch of psychology, the scientific study of behavior and mental processes. For something to be science, or scientifically grounded, a handful of criteria must be met. First, science investigates questions that can be answered using the scientific method. This means when scientists share their findings at the end of the study, they open them up to discussion, scrutiny, and evaluation. A published scientific study includes all the methodological details of who, what, where, when, and how so that other scientists have the opportunity to critique the process and outcomes of the study.

One of the most exciting things about science is that it is not fixed. Science is really just a collection of what is considered truth at one particular time under one particular set of circumstances. This is why we scientists cringe and get very nervous when someone says science has "proven" something. Science does not prove anything because it would be impossible to test a research question under all possible circumstances that would be required to be 100% certain. What we can do, however, is be pretty darn sure. Scientists humbly communicate their openness to being incorrect by using hedged language like, "Our findings *suggest* ..." or, "This strongly *indicates* that X *may* cause Y." By doing this, they're acknowledging that they studied one phenomenon under one set of conditions and so their results should be interpreted as such. So, the next time you're at the mall and someone tries to sell you hand cream that is "scientifically proven" to make your skin softer, run! At best, the cream *may* make your skin baby-soft; at worst, you could end up with a bumpy rash.

It was not *that* long ago that people thought the earth was at the center of the solar system. In fact, those who disagreed were sometimes persecuted for challenging what had been "proven" at the time. Thankfully, great thinkers like Copernicus came along to scrutinize and evaluate this widely accepted idea. They did what any good scientist does; they collected their own data and compared their findings to what was presented as truth at the time. Copernicus shows us just how important it is for a scientific claim to be able to be replicated by others and therefore able to be demonstrated untrue (i.e., falsifiable). To say that science proves anything is to be overly confident. The replicable and falsifiable criteria for science offer a check-and-balance system that is necessary for new knowledge to be generated and old knowledge to be updated.

The origin of the quote "The plural of anecdote is not data" is unknown. And while the origins of the phrase may be indeterminate, the meaning is most certainly not. Anecdotes are stories or accounts that have not and, usually cannot, be evaluated using the scientific method for one reason or another. Sometimes it's because it happened in the presence of only one person, or it cannot be replicated in controlled conditions like a laboratory. For example, someone might say that a feral cat "asked" them to save her kittens by leading them to a muddy pit where the kittens were stuck. This is a story, an anecdote, not scientific evidence of an animal asking a member of another species for help, and so it must be evaluated very carefully.

Anecdotes certainly do have their place in science, which is important to point out, especially for the purposes of this book. While anecdotes are not scientific, they can be the inspiration for scientific investigations where data are collected. Data are observations that have been intentionally collected with some goal in mind. The heartfelt story of the mother cat asking the human to rescue her kittens is not data. On the other hand, if I had a videotape of the encounter, I could observe some different variables and collect some data. A variable is anything that is free to vary or differ. Maybe I review the video footage of the man and the mother cat, and I record the number of times the mother cat walked ahead of the man and then looked back at him and the loudness of her meowing, or I write a detailed description of how the mother cat physically interacted with the man (e.g., brushing against his legs, pawing at his feet each time he stopped walking). These measurements are all data. You'll notice that by collecting data on these different variables, our mother cat story becomes undeniably more credible than the original anecdote.

Taking some inspiration from the hypothetical data about the mother cat and the man, I might now develop a research question that is replicable, falsifiable, and testable using the scientific method. Many animals help each other, but do animals try to get help from humans? This is a very broad question. In science, narrow questions are better than broad ones. How would you respond to the question, "Would you like to take a ride?" Chances are the answer is, "It depends." Where are we going? How long will it take? Is it a car, a rocket ship, or a bucking bronco? The same goes for research questions. Before researchers conduct a study, they must develop clear operational definitions for all of the variables in their research question and plan. By setting the parameters of as many variables as possible, developing an appropriate study is easier, and the process of interpreting and generalizing the results becomes more credible. We'll get to that later on in the chapter.

Keeping the importance of operational definitions in mind, my narrowed research question becomes, "Will mother cats try to get help for their kittens from a human stranger?" Recall there are many variables that I am not controlling for here, such as the age of the mother cats or their kittens, the time of day, or the sex of the human stranger. All of these factors can have an influence on the results. In fact, recently, scientists discovered that the sex of the experimenter significantly affected laboratory rodents' behavior. The animals exhibited more stress behavior and a higher pain tolerance when they were held by a man or by a woman wearing a recently worn man's cotton T-shirt (Sorge et al., 2014). Not surprisingly, this finding sent the scientific community into a bit of a panic but served as a great reminder to consider as many variables as possible when designing and describing research studies.

ANIMAL RESEARCH ETHICS

To know whether mother cats would solicit help from a human stranger, I would need to set up a situation in which real kittens are in danger. Further, it would need to be a situation that was dangerous enough to cause the mothers' protective instincts to kick in. What would that situation need to look like for the kittens, and more importantly, who gets to determine whether it would be okay for me to conduct my study?

Most animal cognition research is conducted by scientists who are affiliated with colleges and universities. Their studies are determined to be ethical or unethical by a review board called the Institutional Animal Care and Use Committee (IACUC). Each school has its own IACUC, which must comprise at least one animal researcher, one nonscientist, one veterinarian, and one community member. This ensures a range of expertise and perspectives. The IACUC is regulated by the American Association for Laboratory Animal Science (AALAS), which passes down and enforces the policies that all animal researchers are held to—whether they work at a university, private facility, or corporation.

Currently, any research involving vertebrates (i.e., animals with a backbone) must seek approval from the IACUC. However, some individual institutions have encouraged or even required IACUC approval for cephalopod (i.e., squid, cuttlefish, octopus) research due to what we have learned about their cognitive abilities. This is just one example of how animal cognition research has informed animal welfare practices.

In order to receive approval for their study, researchers must submit a proposal that includes details about animal care, their planned procedures, as well as how qualified the experimenter is to conduct any special techniques. The proposal is evaluated with a number of goals in mind: (1) efforts to minimize pain and distress to the animal, (2) evidence that the study has not been done before and will contribute new knowledge, and (3) its adherence to federal and veterinary guidelines. Proposals that are not approved by the IACUC can be revised and resubmitted, but the researcher cannot ethically conduct their study until it has been approved.

At this point, you might be asking, "What stops someone from conducting their study if it has been deemed unethical and gets rejected?" One safeguard is our reputation as scientists. Most journals require authors to submit an ethics statement indicating that their study was approved by an IACUC. In all of our combined years, we have never met an animal researcher who was not also an animal lover. If word got out that someone lied in their ethics statement, they would be shunned by their national and international colleagues, in addition to any legal repercussions.

There are many checks and balances to ensure animals involved in research are protected. Those of us who are privileged enough to work with animals appreciate the oversight of the AALAS and IACUC as systems that help us keep our animals safe.

RESEARCH DESIGNS

Part of receiving approval from the IACUC means justifying the procedures that will be used. Once a research question has been developed, the best methodological approach to answering the question must be determined. There are many different kinds of research designs out there, and animal cognition researchers make use of a variety of them. Just like there is a particular tool for a particular job, researchers must match the appropriate research design to their question. The five most common designs are case study, naturalistic observation, correlation, experiment, and comparative experiment. Each of these offers greater credibility and confidence to our claims than anecdotes can. However, each differs in its own way with respect to the kinds of interpretations we can make about our results.

Naturalistic Observation

Like many humans, I (ECW) enjoy watching animals. At this moment, I'm periodically looking up to watch two crows (*Corvus* spp.) foraging across the street in the rain. I do not have a research question; I'm just curiously observing. Sometimes researchers observe animals because they know so little about a particular species, behavior, or ability that they do not know where to begin to ask a research question. This is when inspiration can strike. Other times, a preestablished question can be answered by simply observing, such as when a zookeeper wants to know how an animal is adjusting to its new enclosure. In both of these situations, researchers can conduct a noninvasive (i.e., no intervention by the experimenter) study of behavior called a naturalistic observation.

At face value, the naturalistic observation design might seem too unstructured to be science, but it is not. Meaningful data are collected, patterns are recorded, and explanations about how animals behave can be generated. The main advantage of this design is that it offers researchers a highly realistic, undisturbed look at behavior as it naturally occurs. For example, my (ECW) curiosity about talking African grey parrots (*Psittacus erithacus*) led to a videotaped naturalistic observation about the kinds of things they talk about when they're home alone (Colbert-White, Covington, & Fragaszy, 2011). As it turned out, for Cosmo, the parrot I observed, alone time seemed to be a great opportunity to practice words and sounds and to point out squirrels that might scurry by.

One of the greatest drawbacks of naturalistic observation research is that while there is structured observation, it often lacks a narrow or detailed research question at the onset. For example, an elementary school science class can ask "Does lead sink or float in water?" then go find out. Animal cognition researchers might not know what to ask until they sit, watch, wait, and wait some more—sometimes for days or weeks. Maybe after an extended stay in the forest, one sees a pack of wolves at a river bank and becomes inspired to ask, "Do wolves help each other when they are faced with a natural barrier?" As you can see, this is where the logistics get complicated. In observational research, a question sometimes emerges out of observation, but if the researcher was not able to set up the environment the way they wanted it in advance, observations after the fact cannot help answer the question.

This brings us to a final point about naturalistic observations: There is almost no control over any variables in the environment. When I studied Cosmo, her speech patterns could be influenced by a number of factors that I was not measuring, like the time of day, if she was hungry or satiated, if there was something particularly distracting or stressful outside the window, or the fact that the phone could ring at any time. In some ways, it was a research nightmare, but it was also exactly what I wanted—Cosmo's real behavior under real-life conditions where the environment changed unexpectedly all the time. Of course, it would have been highly problematic if I only conducted one observation session of Cosmo and it happened to be during a thunderstorm. I might have no idea that it was abnormal for her to be silent during my entire observation, but this is why we collect lots of data. The more data we collect, the better able we are to discriminate between normal behavior and outliers.

The naturalistic observation design can provide the inspiration to develop new questions or to answer questions about how behavior happens in its purest form. As long as the goal is not to know if some variable causes a particular behavior, a naturalistic observation might be just the design researchers need to address their research question—whenever their research inspiration strikes!

Case Study

Some of the most famous and informative studies of animal cognition have been through the case study method. Unlike an anecdote, which is just a story, like a snapshot in time, a case study involves intensively and extensively tracking the behavior of a single animal or small group of animals (e.g., Jane Goodall's work with the Kasakela chimpanzee community in Gombe Stream National Park). Behavior is studied over an extended period of time, and conclusions are used to speculate what may be common or possible for the entire species. This makes the case study method particularly attractive for those who study species that are logistically hard to reach or for those who might need decades of intensive interaction or training to answer their research questions. For example, you may be familiar with some famous animal case studies such as Alex the African grey parrot (Pepperberg, 1999) or Koko the gorilla (*Gorilla gorilla*; Patterson & Linden, 1981), both of whom taught us a great deal about animal communication over the course of decades of research.

One of the main advantages of the case study design is the ability to track change or development in behavior over a period of time (Yin, 2013). Maybe you take notice of your horse (*Equus caballus*) nibbling at the stall latch one day. Remembering that the plural of anecdote is not data, you decide to be more systematic and intentional in your observations. You begin collecting data on the types of behaviors she exhibits, the amount of time she spends working on the latch, whether she seems to make more progress on the latch after she watches you use it, and so on. You take careful note of as many variables as possible over the course of multiple days until your horse figures out how to eventually open the latch—much to your amazement and dismay. This in-depth investigation of your horse's behavior over time would be a case study in her problem solving and potential observational learning. If your other horse also learns how to open the stall door, perhaps using his cheek instead of his teeth, and you document his learning in detail, you now have two case studies on horse cognition. You also now have two horses eating your vegetable garden, but that's beside the point.

While you cannot generalize your two horses' behavior to the cognitive abilities of *all* horses, your case studies paint a picture of what horses are capable of. In addition, as implied by exceptional animals like Alex and Koko, case studies allow for more detailed study of individuals with particular talents and abilities, with the potential to discover how they might have developed differently than others of their species. The challenge of not being able to generalize to all members of the species is also what makes the case study a somewhat controversial design (Flyvbjerg, 2006; Meyer, 2001). Anyone with more than one human acquaintance can tell you that each human has different strengths and weaknesses, and the specific abilities of one individual should not be generalized to the abilities of another. The key in the case study design is to differentiate between ability and capability (Kaufman & Kaufman, 2016; Lloyd, 2004). While a case study highlights in great detail the abilities of one particular animal, it provides support for the capabilities of the entire species. When we want to use case study data to show that it is within the realm of possibility for a particular species to do a certain behavior, whether it's learning a made-up language, manufacturing tools, or counting, the term *power study* is sometimes used (Lloyd, 2004; Pepperberg & Funk, 1990; Triana & Pasnak, 1981). A power study is almost like a "proof of concept" and is based on the idea that members of a species have the same general physiological makeup, particularly as it relates to brain structures. Once the basic hardware is in place, something must act on it to create individual differences in abilities. This can be in the form of genetic variation, differing physical environments, nutrition, the presence or absence of cognitive stimulation, or a multitude of other factors. Nonetheless, a power study can help case study researchers claim that if the conditions are right, a parrot's brain is capable of understanding how to count, for example.

While it might seem like case studies are similar to anecdotes, only the former is considered a valid scientific research design. This is because many variables can be controlled (i.e., held constant), and any which cannot be controlled for are carefully recorded and tracked. The case study method is perfectly suited for those who want to trace the slow change or development of a behavior over time (e.g., animal culture), the effects of specialized housing or training (e.g., Washoe the chimpanzee who was raised in a home like a member of a human family; Gardner & Gardner, 1969), or any other rare circumstances. However, the case study does not offer the opportunity to conclude how any of the variables affect behavior. From a case study, I could not argue that being raised in a home like a human child causes a chimpanzee to have better sign language communication skills. *Better* implies a comparison, and in the world of case study, there's only extensive observation of that one animal or small group of animals. For researchers like Jane Goodall, that is perfectly acceptable. Many of the questions she had about chimpanzee behavior could be answered through her intensive observations.

Correlation

What does it mean to say that there is a correlation between two variables? For example, what does it mean that the average group size for primate species (e.g., gorillas tend to live in small groups of 10 or fewer) correlates with the average size of the species' neocortex (i.e., the outer, wrinkly part of the brain that is associated with higher order cognitive abilities; Dunbar, 1992)? Does it mean that living in a larger group causes a species' neocortex to get bigger? Does it mean that having a bigger neocortex causes a species to have larger group sizes? While either or both of these conclusions might seem tempting, neither would be appropriate. This is because there is a strong distinction between correlation and causation in science. Here's another example: When your dog barks at the mail carrier, do his barks *cause* the mail carrier to leave? Your dog might think so, but realistically, the mail carrier leaves because they have completed their job of delivering your mail. Barking and leaving are correlated, but there is no *causal* relationship between those two behaviors.

When two variables are correlated, one can predict the other. Knowing that gorillas live in small groups, I can predict that the gorilla neocortex is probably smaller than a baboon's (*Papio* spp.), since some species of baboon live in groups that can reach as many as 100 individuals. Dunbar's (1992) study correlating primate group size with neocortex size is a fascinating example of a correlation research design. Dunbar was curious about whether two variables related to one another in some predictable way. Correlations can come in two flavors: positive or negative. A positive correlation means that both variables move in the same direction, such as with Dunbar's example of how when group size increases, neocortex size also increases. A negative correlation means that as one variable increases, the other decreases. Two mind-blowing examples of negative correlation come from the world of spider cognition. Spiders (*Araneae* spp.) spin different kinds of webs depending upon many factors such as wind speed, hunger level, and the type of prey they are trying to catch (see Japyassú & Laland, 2017). Research has shown that there is a negative correlation between prey size and the tightness of the web (Watanabe, 2000). Specifically, the smaller the prey, the tighter the web. A tighter web is a more sensitive web, which makes it easier to detect tinier insects. There is also a negative correlation between successful prey capture and the stickiness of the web, whereby the lower the success rate, the stickier the spider spins its next web (Nakata, 2007). This indicates that they track prior failures and use them to hopefully improve their future success rate.

One of the advantages of the correlation design is that the relationship already exists in the world—it's just the researcher's job to uncover it. Spiders have been modifying their webs in predictable ways for who knows how long, but someone had to have the idea to look for a relationship between the different variables. For this reason, correlation studies are a bit lower maintenance than some other designs. By collecting data on only the two variables you predict are correlated, time and costs can be reduced. Correlations are also the best research design to use when it might be unrealistic or even unethical to assess for cause and effect. In order to really know whether having a pet at home during childhood *causes* increases in empathy, we would need to randomly assign a sample of randomly selected families to two groups—one that will have a pet and one that will never have a pet—for 18 years. Though it might be interesting to know the answer to this research question, such a study would never be allowed because it's just not reasonable. Some families who get assigned to the pet group maybe do not want or cannot afford a pet, but they're bound to the rules of the study. What we can do, however, is to collect a large amount of data from as many different adults as possible by asking them how many of their first 18 years of life they owned a pet and then having them take an empathy test. By showing that childhood experience with pets is associated with greater empathy, we can roughly predict someone's empathy if we know about their childhood pet experience, but we cannot say that having a pet for X number of years of childhood will cause Y empathy score. This is why human and nonhuman correlation findings will be worded as "childhood pet ownership is *associated with* increases in empathy," for example—a reflection of the limitation of correlation research.

Experiment

What if we do want to know about whether a particular variable causes a particular behavior? There are three criteria for causation. First, two variables must be established as somehow related. This could come from prior naturalistic observation or correlation research. Second, the relationship between the two variables must be time ordered. That means Variable or Behavior A must always come before Variable or Behavior B. Returning to our dog and mail carrier example from earlier, we can say these first two criteria are definitely met. The third criterion, however, is where correlation and causation are distinct. In order to make conclusions about causation, alternative explanations for the relationship must be ruled out. This is done by controlling for as many potentially relevant factors (i.e., variables) as possible until presumably only one variable, Variable A, remains. That remaining variable can now be tested to determine if it is the cause of the other variable or behavior in question.

Only experiments meet all three of the criteria for causation. Experiments are the most rigorous design because of the many outside variables that must be controlled for in advance of testing the hypothesis. The remaining variable of interest is called the independent variable. The independent variable is the variable that we allow to, well, vary. The dependent variable, then, is the variable or behavior we measure that we think will be affected by the independent variable. For example, the amount of aggression a cow displays might depend on—be affected by—whether the approaching person is familiar or a stranger. To conduct an experiment, we would need to think of all the possible variables that could trigger aggression other than whether the cow knows the human. We would definitely want to dress the familiar and strange human in the same clothing, ensure they are the same sex and similar body type, control for what they say to the cow and how close they get to the cow, and so forth. Once we create two sufficiently identical individuals who only differ by our independent variable of relationship to the cow (familiar or stranger), we can record the cow's behavior and draw conclusions about the causal relationship between human familiarity and aggression in cows.

Experiment research is great if you want to know about cause and effect, but it does not come without drawbacks. By increasing the amount of control we have over the research environment, we inevitably decrease the realism of the entire situation. (That was a great negative correlation example, by the way!) Consider Cosmo the African grey parrot. If I wanted to know whether being alone *causes* her to say particular words or phrases, I would need to control for all of the factors that could contribute to her vocalizations and leave only whether she is alone or not alone as the remaining variable. By the time I've soundproofed the room and gotten rid of all distractions, our parrot would need to be in an empty, windowless laboratory room. This would be so unnatural that she probably would not talk anyway. Plus, I do not care about what parrots say when they're alone in a sterile laboratory room. I want to know about real behavior in the real world. This is why an experiment is not always the right tool for the job.

Comparative Experiment

Because animal cognition's history involves comparisons between humans and nonhumans, we could not leave out the comparative experiment where two or more species' performance is compared on the same task. This design helps answer questions like how domestication influences dogs' reliance on humans. Without directly comparing dogs to wolves on the same task, we'd never know that dogs will look to humans for help sooner, longer, and much more frequently than wolves when they are given an impossible problem to solve or that dogs will follow a human pointing gesture better than wolves (Miklósi et al., 2003). Findings like these teach us valuable lessons about how species evolve different abilities and behaviors. We humans have learned a great deal about the similarities and differences between ourselves and rats, parrots, pigeons, and other apes from this helpful research design.

Do not be fooled. Even though the name has experiment in it, comparative work cannot assess for cause and effect. We can make comparisons. We can say 6-year-old human children outperform chimpanzees on a particular task, but we cannot say being a chimpanzee *causes* them to perform worse. Here's why. If I compare monolingual and trilingual people on a task and the trilingual folks perform better, it could be because they are trilingual, or could it have something to do with other factors or life experiences that are *associated* with being trilingual, such as being more likely to live in one particular geographic region compared to another? Similarly, if I compare hippos to fleas on a visual memory task, you can hopefully see how it rapidly becomes difficult to draw causal conclusions. If the hippo group performs better than the flea group, what questions would you ask? Maybe what the task was or how I scaled it down to flea size, or how I knew the task was equally challenging for both groups, or whether flea vision is the same as hippo vision. I cannot control for all the diverse factors that go along with being a member of one group or another, whether it's languages spoken or species, so I cannot rule out all possible alternatives, which means any results from comparative studies with animals must be presented and interpreted carefully. Despite this, as we will see in the "Human Application" section, comparative experiments can offer insights into how evolution has shaped cognition and behavior between different species, which makes this design a valuable tool.

RESEARCH LOCATIONS

It's fairly hard to imagine a scientist who studies animals without envisioning them climbing through the jungle or grabbing their laptop before it slides off the deck of the boat in rough seas. Some scientists do their research in the field—the animal's natural habitat—while others ditch the bug spray and life jackets and work in the laboratory. Still, others prefer the benefits of hybrid locations like zoos. All of these locations come with their own unique advantages and disadvantages, which we will get to shortly.

Behavior is messy and often changes based on the environment. This is why some choose to work in the laboratory, a controlled system where variables can be held as consistent as possible—for example, the amount of food, socialization, or space the animal has. But, as we saw with my decision to use a naturalistic observation rather than an experiment, there is something to be said for observation of behavior in natural systems. In an ideal world, researchers examine behavior in both the laboratory and the field, then make comparisons in order to construct a more informed understanding of a particular species' abilities.

To be clear, there is some research that can only be done in one setting or another. Work on how pharmacological drugs or neurological changes impact cognition can only be carried out under the controlled conditions of a laboratory. On the other hand, there exist some species, for example, white sharks, which are simply not well suited to laboratory life and testing, despite our curiosities about their behavior. There are also a variety of natural behaviors, such as memory for extended migratory patterns or breeding sites, that can be difficult, if not impossible, to observe in animals under human care. The good news is that technological advancements have made research equipment more powerful, cheaper, and easily portable, which has broadened the types of questions that can be addressed in the field (Kelly, 2018).

When working with animals, hybrid locations like zoos, aquariums, and natural parks can be a one-stop-shop for researchers who want to study a wide range of species (MacDonald & Ritvo, 2016) or species that might be very hard to study in the wild, such as those living in normally inaccessible places like snow leopards (*Panthera uncia*). Though there are some challenges like slower data collection and the risk of distraction by other animals, hybrid locations offer the opportunity to control for more variables than is possible in the wild.

In hybrid locations, researchers can develop basic tests of abilities and investigate various animals' performance. One test, object permanence, involves understanding that a hidden object is not gone. This is often measured by having an animal choose which of two cups has a food reward under it. It's fairly difficult to get a wild chimpanzee to patiently sit for this type of task, but at a zoo, it can be conveniently built into the animal's normal feeding routine. Observing animals living in hybrid spaces allows scientists to learn about their habits—behaviors like social hierarchies, preferred hunting techniques, and care of offspring—which can then be used for many reasons. In the case of the snow leopard, this might involve convincing the powers that be to preserve their unique alpine mountain habitat (Jackson & Hunter, 1996).

Given the differences of these three types of environments, information gleaned about animals' cognitive abilities should always be considered within the context of the location in which the behavior was studied. Strictly speaking, in terms of pure cognitive abilities, the location of a research study should not matter—an ape is an ape, regardless of whether he is in the wild, a zoo, or a research laboratory. Realistically, that obviously is not always the case. There are features inherent to the research location that affect cognition and cause differences to emerge. For example, responses to humans can differ drastically between the wild and the laboratory. An animal born and raised in human care will generally not show the level of fear of humans that one born in the wild likely would. These encultured animals, as they are called, differ from their wild counterparts in that they have usually been raised by humans and immersed in the human experience. This involves human physical contact, being spoken to, and being exposed to human behavior, body gestures, culture, and technology. Oftentimes they have also participated extensively in a variety of cognitive and linguistic studies. Koko the gorilla and Alex the African grey parrot are great examples of encultured animals. Koko and Alex experienced life far different than their wild counterparts, yet they remain anatomically the same species (Patterson & Linden, 1981; Pepperberg, 1999). Encultured animals like them show us the cognitive potential of the species as a whole, even if all members of that species never reach that potential (Kaufman & Kaufman, 2016).

DATA COLLECTION

Data are the evidence researchers collect that help them answer their research questions. Data can come in two forms. Qualitative data involve detailed records of how a behavior happens, such as documenting in words the process of a mother dolphin teaching her calf how to use sponges to stir up prey on the ocean floor (Krützen et al., 2005). Quantitative data, on the other hand, are numerical representations of behavior, such as how many times a mouse runs a maze or how long it takes an orangutan to solve a puzzle (Galsworthy et al., 2005; Keller & Delong, 2016). Any variable that can be measured can typically be quantified. Some examples include accuracy (e.g., 9 out of 10 correct responses, so 90%), duration (e.g., how long it takes a mouse to complete a maze), latency (e.g., how long it takes until a startled meerkat emerges from her burrow), or frequency (e.g., how many stress behaviors a dog emits while trying to solve a puzzle).

Here's a question: What does a stress behavior look like in a dog? This is where operational definitions come in handy once again. In dogs, yawning and shaking off can be signs of stress. Therefore, "stress behavior" might be operationally defined as specifically yawning or shaking off. To be even clearer, the researcher could also write out specific definitions for what yawning and shaking off look like. I might define a yawn as *opening the mouth wide, teeth become exposed, air is inhaled then exhaled, lasts more than 2 seconds*. While it might seem overly detailed, the clearer, the better!

Any list of a species' typical daily activities or those that are specifically being counted is called an ethogram. Developing an ethogram for a particular animal or group of animals helps to record behavior for both a basic understanding of the species and comparison when the situation changes, like if the dog emits more stress behaviors when there's a human in the room versus not. Researchers interested in the effects of humans on wild animal behavior often use ethograms. To document how nearby development impacts the behavior of a population of chipmunks, they would need an ethogram of natural chipmunk (*Marmotini* spp.) behavior before the development started. If that ethogram showed that the chipmunks normally spent 10% of their time on alert for predators and now they spend 30% of their time that way, one could reasonably conclude that the development was having an impact on them because their natural vigilance behavior has increased threefold.

Ethograms can be broad and catalog all the behaviors of a species (Shettleworth, 1975, has a great one for hamsters, *Mesocricetus auratus*, which includes definitions for species-typical behaviors such as scent marking, face washing, scrabbling, and hoarding) or narrow and apply specifically to behaviors being measured for a study. For example, it would be possible to create an ethogram of only aggressive behaviors, but the trick would be to make sure everything is specifically defined. While yawning is a behavior that most people recognize, describing an animal's behavior as "fighting" really is not very helpful—it's too vague and could be defined differently by different people, similar to "stress behavior." On the other hand, *rearing on hind legs, biting, swiping with paws, and charging at an opponent* would be a much more informative way to operationally define fighting.

ASSESSING RELIABILITY

Because animal cognition is often studied by observing the outward behavior of an animal, issues related to determining an observer's reliability (i.e., accuracy, consistency) are particularly important to the integrity of a study. However, if you've ever tried to mediate the behavior of small children, you know how many sides there can be to the same story.

"Mom, Jacob pushed me!"

"He hit me first! I got the last muffin, and he tried to grab it and hit me!"

Anyone feel like coming to dinner at my (AK) house? It's not hard to see why your hand is not raised. Even if you had been carefully observing what happened between my sons, you might not be confident in your determination of what actually occurred. Behavior happens fast. It happens loudly. It happens in three dimensions. What looks like a push from one angle might look like a hit from another, and two people recording my sons' behaviors might write down different observations.

As a result, reliability measurement is vital. In order to come to conclusions based on what behavioral observations were made, researchers cannot be the only judge of what they see (Cohen, 1960; Kazdin, 1977). Ensuring reliability can be done in various ways. In the most fortunate of circumstances, two or more observers watch an animal's behavior, looking for the previously agreed upon target behaviors, such as whether a dog chooses the left or right container when given a choice (e.g., Colbert-White, Tullis, Andresen,

Parker, & Patterson, 2018). The two observers then compare notes, and if high interobserver agreement is reached, we would say there is high reliability for what they have seen (Cicchetti & Sparrow, 1981; Fleiss & Cohen, 1973; Suen & Lee, 1985).

Movements and actions can be subtle, and observers may miss behaviors or interpret them incorrectly. If both observers are not careful, do not have the same observation line of sight, or have not been trained extensively together on the operational definitions for the behaviors they are measuring, strong reliability can be hard to obtain. In addition, there is more room for accidental bias than scientists like to admit. For these and other reasons, assessing reliability is an important check and balance on the research process.

INTERPRETING FINDINGS

Beginning with a review of the scientific method, we explored how research questions can be developed from a variety of sources, paying careful attention to differentiating data from anecdotes. From there, we considered the importance of designing methods that are ethically sound and appropriate for the research question, while taking into account the location where data collection will occur. Then, we discussed what kind of data can be collected and how to do it in a way that maintains integrity. According to the scientific method, after collecting data, it must be analyzed and interpreted. We finish this chapter by revisiting important lessons first learned by pioneers of the field in the early 20th century. These lessons are the risks of anthropomorphizing and overstepping conclusions.

The number of Internet animal videos we (AK and ECW) ruin for our friends and family is staggering. Sure, we like stories of animals saving other creatures, performing brilliant feats of cognition, or acting humanlike, but chances are, a dog or cat rescuing a bird from drowning is probably looking for lunch, not a new best friend. And maybe the crow (*Corvus* spp.) that is "sledding" down a snowy rooftop on a piece of plastic is happening purely by accident, much to the crow's frustration. As we pointed out in the previous chapter, most people, ourselves included, tend to anthropomorphize, or attribute humanlike behaviors and traits to animals. While there are benefits to anthropomorphism, such as being open-minded to the possibility of more advanced behaviors in animals or offering a starting point for

CHAPTER 2 THEORETICAL AND METHODOLOGICAL APPROACHES

considering why an animal might behave a particular way, prominent animal cognition researcher Sara Shettleworth (2010) identifies one way that anthropomorphism can become problematic:

> Although the extent of human–animal cognitive similarity is undoubtedly a key issue for comparative psychology, it sometimes seems the agenda is to support anthropomorphic interpretations rather than to pit them experimentally against well-defined alternatives. The enthusiasm of the popular and even scientific press for clever animal stories nourishes this tendency: killjoy explanations are less likely to make headlines than stories about how octopi or birds are unexpectedly human-like. (p. 478)

Shettleworth's quote refers to a more human-centered anthropomorphism, which is highly uninformed and does not separate the person from the animal. Many species of animals are not only solitary but also aggressive. People have trouble imagining a desire to live a solitary life, so they may say a tiger is "lonely," when it would actually not naturally seek out companions (Timberlake, 1997). Likewise, an animal that does not appear to have a concept of past or future (Roberts, 2002) may not be aware of its habitat size at the zoo or other features that we humans would notice immediately if we were in their situation.

A video of an octopus or bird acting in a human way immediately confirms what someone might expect to see because it looks familiar and fits their worldview of how *they* would respond. As a result, they might see no need for further investigation into the animal's natural history, which is why human-centered anthropomorphism is uninformed and problematic. One of the reasons animal cognition researchers follow strict protocols regarding interobserver reliability (Kaufman & Rosenthal, 2009), research design (Herman, 2010; Highfill & Kuczaj, 2010; Trestman, 2015), and evaluation of their work by colleagues in the field is to address the easy tendency to fall into the trap of anthropomorphic bias. On the other hand, not all anthropomorphism is bad. When researchers take an animal-centric approach, they put themselves "in the shoes of" the animals they are studying after they have learned about that species' biological predispositions, social organization, predators, and habitat (Timberlake, 1997). Animal-centered anthropomorphism can be a useful tool during the data interpretation process.

When we think of an early pioneer who exemplified both the methodological good and bad of studying animals' cognitive abilities, George Romanes comes to mind. We first introduced Romanes in Chapter 1, Historical Perspective on Animal Cognition, but return to him briefly to segue to one of animal cognition's earliest data interpretation mantras.

Romanes often used anecdotes and anthropomorphism. His 1878 book, *Animal Intelligence*, was a collection of stories from friends, pet owners, naturalists, and zookeepers. Romanes did a great job of providing lots of evidence to build up his arguments for complex experiences like emotion, creativity, and problem solving in other species, but the many uncontrolled factors associated with anecdotes made his claims challenging to replicate. Further, because many of the anecdotes in Romanes' book came from individuals who were not scientists, they rarely had a second observer to corroborate their accounts.

Shortly after Romanes published *Animal Intelligence*, pioneer C. Lloyd Morgan (1894) made it clear that though he appreciated the anecdotal approach, it was not science (Wasserman, 1993). At the time, Morgan had been refining methods for assessing animal intelligence by studying newly hatched chicks that had been isolated from their mothers. The designs of his experiments and the precautions he took to keep his methods and interpretations as rigorous as possible explain why he wanted to ensure the field was heading in the right direction.

As part of his charge to his contemporaries, Morgan (1894) laid down the following precaution:

> In no case may we interpret an action as the outcome of the exercise of a higher psychical faculty, if it can be interpreted as the outcome of the exercise of one which stands lower on the psychological scale. (p. 53)

That is to say, in interpreting animal behavior, observers should work from a place of skeptical simplicity. Rather than arriving at the conclusion, for example, that ants are able to tell time because they arrive at picnic tables at 12 p.m. each day, Morgan would encourage the observer to consider what simpler underlying process could be happening. Arriving "on time" for an activity is a highly complex, abstract skill, which would require a concept of numbers and time, an understanding of future, an ability to plan, and so on. On the other hand, we could also explain arriving at 12 p.m. as ants learning

CHAPTER 2 THEORETICAL AND METHODOLOGICAL APPROACHES

where a consistent food source is located and it taking them a certain amount of time to arrive at the picnic tables each morning based on where their nest is. The latter interpretation reduces the number of complex cognitive abilities necessary to produce the same behavior. Morgan cautioned that before jumping to the "top" of the cognitive complexity hierarchy to interpret behavior, alternative, simpler explanations should be ruled out first so as to avoid bias and error (Costall, 1993). This is called parsimony, or Morgan's Canon, as it came to be known. Along with abiding by the criteria of using the scientific method, being replicable, and falsifiable, parsimony is a fourth criterion for science. Morgan's Canon became the standard by which animal cognition researchers evaluated their data for decades, and by and large it still is.

Unfortunately, though his goal was neither to say animals are incapable of cognitively complex behaviors (Rollin, 1986) nor that anthropomorphism should always be rejected, Morgan's Canon was used to conveniently disregard the study of animal minds. If I can boil down the ability to tell time to an instinct or conditioned response that requires no discussion of mental states, I can develop more objective laws of behavior—which was just what some psychologists at the turn of the 20th century wanted. The behaviorists used Morgan's call for parsimonious interpretation as a way to silence any discussion of animals' mental abilities for decades. And so, Morgan's Canon, considered by some to be the most widely quoted statement in the history of psychology (de Waal, 2001), essentially returned animals to machines and helped behaviorism become a dominant psychological perspective for more than four decades.

In a fascinating plot twist, Morgan seemed to know that his famous statement might one day be misinterpreted on a large scale. In a revised edition of his 1894 book, he offered a clarifying sentence right after his famous canon:

> To this, however, it should be added, lest the range of the principle be misunderstood, that the canon by no means excludes the interpretation of a particular activity in terms of the higher processes if we already have independent evidence of the occurrences of these higher processes in the animal under observation. (p. 59)

This revision helped animal cognition come full circle, back to the goals of those who had originally pioneered the field. Nonetheless, generations of animal cognition researchers were trained to interpret findings using Morgan's original canon. Consequently, even into the 21st century, some argue it remains a barrier to accurately studying animals' cognitive abilities (e.g., Fitzpatrick, 2008).

Given the many ways in which behavior and the environment can vary, animal cognition researchers must be strong critical thinkers. We arm ourselves with an evolving tool kit of diverse methods, mantras, and considerations in order to accurately represent animals' abilities. Arguably, the ability to think and interpret data from multiple perspectives is our superpower, but it was not always that way. Much of our research methods' tool kit was developed through trial and error over its first century. Still today, we understand that the burden of proof is on us to provide compelling evidence, and strong methodology is critical to doing so.

ANIMAL SPOTLIGHT: CLEVER HANS

Horses, in general, are not often studied for their cognitive abilities—because of their historical use for transportation and work, it's often assumed their cognitive powers are not particularly impressive (Hanggi, 2005). We do know that they have some ability to generalize stimuli. This means a horse that is trained to always touch a large circle when given a choice between shapes will likely touch a medium-sized circle it's never seen before in a new set of shape discrimination trials. Applying that experiment to real life, the ability to flexibly generalize across situations might be important for a horse who is grazing on a variety of grasses or one who needs to perform the same behaviors regardless of the skill level of multiple riders (Hanggi, 2005). At least one study has shown that horses are able to categorize objects via rules (i.e., small versus large) and maintain those rules in long-term memory. The horses in this particular study were able to remember the categorization rules they had learned and even apply them to novel tests 7 years later (Hanggi & Ingersoll, 2009). At a higher level, there is some initial evidence for the ability of horses to understand what a human caretaker does or does not know—an ability that will be discussed later in this book. One study showed that if a horse—but not his caretaker—saw food hidden, the horse's behavior changed so that it more often touched the caregiver and looked in the direction of the hidden food (Ringhofer

& Yamamoto, 2017). Interestingly, the jury is still out on observational learning for this fairly social species. Wild horses generally live in large groups, so we might suspect that it's fairly common for one individual to learn from another simply by watching them. Surprisingly, scientists have found that horses do not generally look to each other for more information or learn from each other, with the possible exception of subordinate animals learning from dominate ones (Brubaker & Udell, 2016).

Chances are, if you asked any of the researchers from these studies to name the most famous horse in all of animal cognition, you'd get the same answer. This is because the story of Clever Hans is widely considered to be one of the first animal case studies, dating back more than a century. Clever Hans, a horse owned by math teacher Wilhelm von Osten, was reported to be able to count and communicate in German (Pfungst, Stumpf, & Rahn, 1911). Hans could answer math problems, knew the calendar and the clock, and could answer questions and communicate in words! Hans would simply stamp his hoof, providing numerical answers or numbers corresponding to letters of the alphabet on a blackboard.

And Hans was good at it. Very good. So good that it took esteemed scientists 3 years of carefully studying Hans to figure out how he always seemed to get the correct answer. One day, one of the scientists asked, "What would happen if no one else in the room could see the question except Hans?" Ah-ha! While Hans got 98% of the questions correct if the human questioner knew the answer, Hans was only right 8% of the time if the human questioner either did not know the answer or if Hans could not see him (Pfungst et al., 1911). As it turned out, Hans was not "clever" at math or telling time; he had an uncanny ability to detect subtle physical cues from the body language and microexpressions of the people who were present during his testing. Paul Ekman and others later studied this phenomenon in humans and concluded these cues are both unintentional and uncontrollable in almost every situation (Ekman, 1970; Haggard & Isaacs, 1966).

From Clever Hans arose an age of skepticism that has both plagued and enhanced the study of animal cognition. Animal cognition findings are subject to scrutiny beyond that of human research, assuming they're not dismissed outright. It makes our research stronger and challenges us to create more rigorous research designs, but it can be frustrating as well.

Training yourself to be skeptical is a hard lesson to learn; it's certainly not instinctual. In a more recent case, researchers once again fell victim to the Clever Hans Effect when they sought to test the ability of elephants (*Elephas maximus*) to discriminate between quantities of items (Irie & Hasegawa, 2012; Irie-Sugimoto, Kobayashi, Sato, & Hasegawa, 2009). The elephants were shown food items dropped in one of two buckets, and then they were allowed to approach and eat from a bucket of their choice. An elephant who is able to track the number of items going into the buckets should choose the bucket with the most food. Indeed, the elephants in Irie-Sugimoto et al.'s study ate from the bucket with more food. But they did more. They were able to detect the difference and make appropriate choices even when the quantities in the two buckets differed by only one or two pieces of food, something humans struggle with as the quantities get larger.

Although we cannot be sure, it's possible this was a Clever Hans-like situation. In Irie-Sugimoto et al.'s procedure, the experimenter hid their face and did not make eye contact with the elephants during or after placing the food items in the buckets. However, when another research team added specific controls for the Clever Hans effect to Irie-Sugimoto et al.'s study and replicated the procedure, it appeared, to quote the authors, that their clever elephants belonged "right back with the herd" (Perdue, Talbot, Stone, & Beran, 2012, p. 995). In Perdue et al.'s procedure, the containers into which the food was dropped were behind a handler who faced the elephant and wore noise-canceling headphones. A second person, the experimenter, baited the containers and then left. As a result, the keeper—who was the only person who interacted with the elephant—could not see or hear which container contained more food. These modifications prevented the elephants from making eye contact with anyone who knew which container held more food. Under these conditions, the elephants were far less successful and performed more like other elephants in the past (Perdue et al., 2012).

A century later, Clever Hans continues to reinforce important lessons for researchers. While Hans was not the math whiz, we were fooled to believe he was; he taught us just how closely other species pay attention to us and how tiny, predictable changes in our behavior can be attached to meaning for them. So, the next time you think your cat can read your mind because she "always knows" when you're getting ready to leave the house, stop and remember Morgan's Canon. What behaviors predict your departure? The jingle of your keys might be an easy one to come up with, but what about something less obvious, like unplugging your cell phone in the kitchen? Your cat does not have extrasensory perception, but she is really good at reading your body language and other predictable cues. For animal researchers who try to control the environment and be as objective as possible, this reminder makes our job all the more challenging and exciting!

HUMAN APPLICATION: AN ORIENTATION TO EVOLUTIONARY BIOLOGY

Why should the field of animal cognition exist? Sure, we love our animals, but who cares what they think or of what they are capable? Or more specifically, who cares if our methods lead us down the wrong path? Mother cats ask humans for help when they know their kittens are in danger. What does it matter if that's actually true or if I just want to believe it's true? Well, if we have conducted methodologically sound research and we determine that cats actually will seek out help from other species, we could think about the ways in which being willing to solicit help from other species was advantageous for cats' ancestors. Further, we can then make comparisons to human behavior. When we systematically study animal cognition, we get clues about how we, as humans, have evolved our own advanced abilities, such as altruism and social helping behavior.

To begin to understand how human abilities evolved, it's sometimes helpful to have a basis for comparison, and animals can serve that purpose (Haun, Jordan, Vallortigara, & Clayton, 2010; Herrmann, Call, Hernández-Lloreda, Hare, & Tomasello, 2007; Trestman, 2015). Studying nonhuman animals provides new perspectives on the different ways that cognition manifests itself and can provide insights into the evolutionary processes that led to our abilities. Darwin's theory of evolution showed at a basic level that when it comes to behavior and cognition, we are different more in degree rather than kind. That is to say, behaviors that we have considered distinctly different from other species, setting us apart in kind, could actually be observed in other species, albeit to a different degree or complexity. For instance, chimpanzees, elephants, humans, and crows can all make and use tools, something we used to think made humans unique.

If we know that tools are common among these four species, did the ability evolve independently, or did all of these species inherit it from a recent common ancestor who also made and used tools? The field of evolutionary biology, and specifically phylogenetics, can help answer this question as well as others related to the uniqueness of humans' cognitive abilities. One of the tools that evolutionary biologists use is a phylogenetic tree, something you may have seen in an introductory biology textbook. Phylogenetic trees are helpful tools that visualize the evolutionary relationships among different species so that similarities and differences in physical, behavioral, and cognitive traits and abilities can be investigated through an evolutionary context (de Queiroz & Wimberger, 1993).

If different species share a particular ability, it means it evolved in one of two ways. We've used tool use as an example in Figure 2.1.

The first way involves the two species inheriting the ability from a common ancestor, which would make the ability homologous. The species have the ability because the ancestor they most recently shared also had it, and they have both retained the ability. Looking at Figure 2.1, this is likely the case for humans and chimpanzees with respect to many similar cognitive abilities, such as tools. On the other hand, sometimes different species share a cognitive ability not because they inherited it from a common ancestor but because they

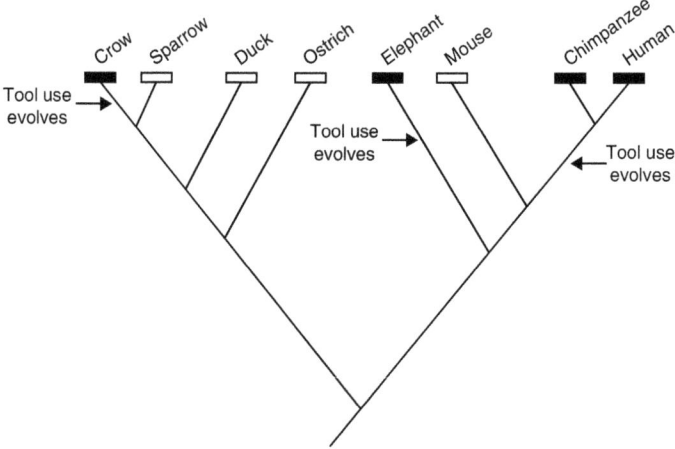

FIGURE 2.1 Simplified phylogenetic tree demonstrating evolutionary relationships among species.

CHAPTER 2 THEORETICAL AND METHODOLOGICAL APPROACHES

arrived at the same ability by evolving it independently. In convergent evolution, two species both have an analogous ability, where the ability is common among multiple species but not because a recent common ancestor passed it down. Given the phylogenetic distance among primates, elephants, and crows, the figure shows multiple times where tool use evolved over millions of years of species change and adaptation. A lot can happen during the 5 or so million years between when humans and chimpanzees shared a common ancestor, but a whole lot more can happen in the more than 100 million or 300 million years separating humans from elephants and birds, respectively! Thus, it makes sense that when species are highly evolutionarily different, common abilities might still pop up independently.

The question then becomes, what causes two totally different species, like crows and humans, complete with different anatomy, physiology, neurology, habitats, and other factors, to both evolve the similar ability to make and use tools? Evolutionary biologists believe that ecological pressures are key here. Somehow, despite their many genetic and environmental differences, both humans and crows shared a need to use objects in their environment as extensions of themselves to solve problems. The materials both species use as tools may be different, the types of problems we encounter and use tools to solve may be more different, but the underlying cognitive ability required to see an object in the environment and harness and use it is similar. At the end of the day, individuals who are more successful at surviving and reproducing will pass on their (more successful) abilities. The ancestral crow and ancestral human who separately began manufacturing and using tools paved the way for those impressive abilities we see in modern versions of those species.

Rather than reaching to anthropomorphism to interpret findings, evolutionary biology can offer helpful food for thought when trying to make sense of humanlike cognitive abilities in animals. Kanzi the bonobo, Alex the parrot, and Phoenix and Akeakamai the dolphins (*Tursiops truncatus*)—just to name a few—obviously do not all share a recent common ancestor. Yet, they seem to have similar cognitive abilities to humans related to the abstract thinking associated with language (Pepperberg, 2001; Trestman, 2015). We could frame this by placing humans at the top of some uniquely human cognitive hierarchy and seeing how other species compare to us. Or we could frame cognitive abilities among all species, humans included, as a result of convergent evolution that differs in degree, not kind. When we do this, we can think about the many different circumstances under which independent evolution of similar cognition might have occurred, and why. Were there particular social, physiological, or environmental pressures that would make it advantageous for the ability to evolve more than once in the animal kingdom?

One final idea merits mention is the concept of deep homologies (Shubin, Tabin, & Carroll, 1997, 2009). While crows and humans likely evolved the ability to make and use tools independently via convergent evolution, there are some basic abilities required for tool making and use that are actually homologous at some point further back in evolutionary time. In deep homologies, abilities or traits that appear to have been independently acquired may actually share genes inherited from a common ancestor that provide the foundation for similarities seen among a diverse group of species. These genes can be "turned on," "turned off," or modified via evolutionary pressures in different species to create the variety of traits we see. Deep homologies for cognition might involve sets of genes responsible for aspects of language or cognitive processes like abstract representation, syntactical ability, numerical ability, understanding point of view, or sentence creation (Chomsky, 2007; de Waal & Ferrari, 2010; Rutishauser & Moline, 2005; Scharff & Petri, 2011). In order to make and use tools, it's necessary to be able to learn from the consequences of actions (e.g., knocking a rock on a nut opens it) and to be exploratory enough to even interact with the rock in the first place, for example. Associative learning and neophilia (attraction to new things), then, might be deep homologies for crows, elephants, humans, and chimpanzees. Taking this approach, where certain building block abilities are added upon differently by different species depending upon their evolutionary needs, would explain why scientists have found species that share some humanlike advanced cognitive abilities and other species that only share the foundational cognitive abilities with us.

Understanding these different evolutionary pathways using a phylogenetic approach helps us to determine how and why certain cognitive abilities evolved, as well as the likelihood that certain advanced cognitive abilities are uniquely human (Haun et al., 2010; Trestman, 2015; Vonk & Povinelli, 2006). In doing so, we have come to learn that Darwin was right; when it comes to just how special or unique humans are, the answer seems to be not very. And we have evolutionary biology to thank for this humbling reminder.

YOUR TURN!

Most people have never spent more than 5 minutes closely observing an animal without interruption or distraction. This first research-at-home idea involves one of the most important qualities an animal cognition researcher needs in order to be successful: patience. To complete this ministudy, you'll need a stopwatch, paper, and a writing utensil. Cell phones can be distracting, so a real stopwatch is encouraged, if possible. Find an animal to observe—at the zoo, your home, in your backyard, or any other place that you can observe one (or a very small number) of animals. To avoid distracting the animal, try to position yourself where you can see them clearly, but you are as nonthreatening and uninteresting as possible.

Create two columns on your paper. These columns will be for behaviors and operational definitions. When you're ready, set your timer for 15 minutes, and start watching. As you observe, make note of behaviors you see, and jot down brief definitions of what those behaviors look like. Your definitions should be clear enough for someone to pick up your ethogram and use it to observe and identify the ethogram's behaviors without your assistance.

The overall goal is to practice careful, structured observation. Remember the importance of detail in your behavior labels as well. "Walking" is a behavior, but so are "walking with tail pointed straight up" and "walking while sniffing the ground." The more detail you put into your behaviors and operational definitions, the clearer a picture you'll have for yourself later.

Here's where the patience really comes into play. What if your animal sleeps the entire time? Those are real data! What if your animal runs, hops, or slithers away on its own? Also real data. What if you accidentally sneeze and scare your animal away or make them approach you? Also real data. If any of these or other unexpected situations (or frustrations) arise, stop, or regroup, maybe select a different species or time of day, reset the timer, and start again.

At the end of 15 minutes, take a break, and flesh out your ethogram. You'll want at least 5 to 10 behaviors with clear operational definitions. Once your ethogram is finished, you can now try a number of different variations. You could observe for another 15 minutes and make tally marks each time the animal does a behavior on your ethogram, which will start to translate your observations into quantitative data. You could give your ethogram to someone and have them use it to identify behaviors, which will test out how clear your operational definitions are. If it's possible, you could try an intervention—placing a novel object nearby, playing music, or allowing the animal to see you—and observe for another 15 minutes. Maybe like a zookeeper wanting to know how an animal is adjusting to its new enclosure, perhaps you now have a research question to pursue, such as how the deer in your backyard respond to your new garden gnome. Do they engage in more instances of "ears erect" behavior and "cautious walking" now compared to before the gnome arrived?

Once you've figured out the basics of ethogram building, behavior tallying, operational definitions, and patient watching, the animal observation world is your oyster! Our only request is that you avoid creating interventions that you predict would be intentionally stressful for the animal, like taking the vacuum cleaner out of the closet, unless it's something that is a part of your daily routine anyway. That is to say, rather than taking the vacuum out of the closet just for the purposes of observation, time your observation to occur the next time you plan to vacuum. Who knows, maybe the quantitative behavioral changes you see in your cat's behavior are so drastic you decide to move to a silent push vacuum instead of a loud electric one. By engaging in close observation, much can be learned about the animal experience.

REFERENCES

Brubaker, L., & Udell, M. A. R. (2016). Cognition and learning in horses (*Equus caballus*): What we know and why we should ask more. *Behavioural Processes, 126*, 121–131. doi:10.1016/j.beproc.2016.03.017

Chomsky, N. (2007). Biolinguistic explorations: Design, development, evolution. *International Journal of Philosopical Studies, 15*(1), 1–21. doi:10.1080/09672550601143078

Cicchetti, D. V., & Sparrow, S. S. (1981). Developing criteria for establishing interrater reliability of specific items: Applications to assessment of adaptive behavior. *American Journal of Mental Deficiency, 86*, 127–137.

Cohen, J. (1960). A coefficient of agreement for nominal scales. *Educational and Psychological Measurement, 20*(1), 37–46. doi:10.1177/001316446002000104

Colbert-White, E. N., Covington, M. A., & Fragaszy, D. M. (2011). Social context influences the vocalizations of a home-raised African Grey parrot (*Psittacus erithacus erithacus*). *Journal of Comparative Psychology, 125*(2), 175. doi:10.1037/a0022097

Colbert-White, E. N., Tullis, A., Andresen, D. R., Parker, K. M., & Patterson, K. E. (2018). Can dogs use vocal intonation as a social referencing cue in an object choice task? *Animal Cognition, 21*, 253–265. doi:10.1007/s10071-018-1163-5

Costall, A. (1993). How Lloyd Morgan's Canon backfired. *Journal of the History of the Behavioral Sciences, 29*, 113–122. doi:10.1002/1520-6696(199304)29:2<113::AID-JHBS2300290203>3.0.CO;2-G

de Queiroz, A., & Wimberger, P. H. (1993). The usefulness of behavior for phylogeny estimation: Levels of homoplasy in behavioral and morphological characters. *Evolution, 47*(1), 46–60. doi:10.1111/j.1558-5646.1993.tb01198.x

de Waal, F. B. M. (2001). *The ape the sushi master: Cultural reflections by a primatologist.* New York, NY: Basic Books.

de Waal, F. B. M., & Ferrari, P. F. (2010). Towards a bottom-up perspective on animal and human cognition. *Trends in Cognitive Sciences, 14*(5), 201–207. doi:10.1016/j.tics.2010.03.003

Dunbar, R. I. (1992). Neocortex size as a constraint on group size in primates. *Journal of Human Evolution, 22*(6), 469–493. doi:10.1016/0047-2484(92)90081-J

Ekman, P. (1970). Universal facial expressions of emotion. *California Mental Health Research Digest, 8*(4), 151–158.

Fitzpatrick, S. (2008). Doing away with Morgan's Canon. *Mind & Language, 23*, 224–246. doi:10.1111/j.1468-0017.2007.00338.x

Fleiss, J. L., & Cohen, J. (1973). The equivalence of weighted kappa and the intraclass correlation coefficient as measures of reliability. *Educational and Psychological Measurement, 33*(3), 613–619. doi:10.1177/001316447303300309

Flyvbjerg, B. (2006). Five misunderstandings about case-study research. *Qualitative Inquiry, 12*(2), 219–245. doi:10.1177/1077800405284363

Galsworthy, M. J., Paya-Cano, J. L., Liu, L., Monleón, S., Gregoryan, G., Fernandes, C., … Plomin, R. (2005). Assessing reliability, heritability and general cognitive ability in a battery of cognitive tasks for laboratory mice. *Behavior Genetics, 35*(5), 675–692. doi:10.1007/s10519-005-3423-9

Gardner, R. A., & Gardner, B. T. (1969). Teaching sign language to a chimpanzee. *Science, 165*(894), 664–672. doi:10.1126/science.165.3894.664

Haggard, E. A., & Isaacs, K. S. (1966). Micromomentary facial expressions as indicators of ego mechanisms in psychotherapy. In *Methods of Research in Psychotherapy* (pp. 154–165). Boston, MA: Springer US. doi:10.1007/978-1-4684-6045-2_14

Hanggi, E. B. (2005). The thinking horse: Cognition and perception reviewed. *AAEP Proceedings, 51*, 246–255.

Hanggi, E. B., & Ingersoll, J. F. (2009). Long-term memory for categories and concepts in horses (*Equus caballus*). *Animal Cognition, 12*(3), 451–462. doi:10.1007/s10071-008-0205-9

Haun, D. B. M., Jordan, F. M., Vallortigara, G., & Clayton, N. S. (2010). Origins of spatial, temporal and numerical cognition: Insights from comparative psychology. *Trends in Cognitive Sciences, 14*(12), 552–560. doi:10.1016/j.tics.2010.09.006

Herman, L. M. (2010). What laboratory research has told us about dolphin cognition. *International Journal of Comparative Psychology, 23*(3), 310–330. Retrieved from https://escholarship.org/uc/uclapsych_ijcp

Herrmann, E., Call, J., Hernández-Lloreda, M. V., Hare, B. A., & Tomasello, M. (2007). Humans have evolved specialized skills of social cognition: The cultural intelligence hypothesis. *Science, 317*(5843), 1360–1366. doi:10.1126/science.1146282

Highfill, L. E., & Kuczaj, S. A. (2010). How studies of wild and captive dolphins contribute to our understanding of individual differences and personality. *International Journal of Comparative Psychology, 23*(3), 269–277.

Irie, N., & Hasegawa, T. (2012). Summation by Asian Elephants (*Elephas maximus*). *Behavioral Sciences, 2*(2), 50–56. doi:10.3390/bs2020050

Irie-Sugimoto, N., Kobayashi, T., Sato, T., & Hasegawa, T. (2009). Relative quantity judgment by Asian elephants (*Elephas maximus*). *Animal Cognition, 12*(1), 193–199. doi:10.1007/s10071-008-0185-9

Jackson, R., & Hunter, D. O. (1996). *Snow leopard survey and conservation handbook*. Seattle, WA: International Snow Leopard Trust.

Japyassú, H. F., & Laland, K. N. (2017). Extended spider cognition. *Animal Cognition, 20*(3), 375–395. doi:10.1007/s10071-017-1069-7

Kaufman, A. B., & Rosenthal, R. (2009). Can you believe my eyes? The importance of interobserver reliability statistics in observations of animal behaviour. *Animal Behaviour, 78*(6), 1487–1491. doi:10.1016/j.anbehav.2009.09.014

Kaufman, J. C., & Kaufman, A. B. (2016). Capacity, potential, and ability: Integrating different approaches to studying animal vs human creative processes. *RUDN Journal of Psychology and Pedagogics, 4*, 29–36. doi:10.22363/2313-1683-2016-4-29-36

Kazdin, A. E. (1977). Artifact, bias, and complexity of assessment: The Abcs of reliability. *Journal of Applied Behavior Analysis, 10*(1), 141–150. doi:10.1901/jaba.1977.10-141

Keller, A. M., & Delong, C. M. (2016). Orangutans (*Pongo pygmaeus pygmaeus*) and children (*Homo sapiens*) use stick tools in a puzzle box task involving semantic prospection. *International Journal of Comparative Psychology, 29*, 1–25. Retrieved from https://escholarship.org/uc/uclapsych_ijcp.

Kelly, D. M. (2018). Animal cognition: An integrative approach. *Animal Cognition, 21*, 1–2. doi:10.1007/s10071-017-1151-1

Krützen, M., Mann, J., Heithaus, M. R., Connor, R. C., Bejder, L., Sherman, W. B., & Sherwin, W. B. (2005). Cultural transmission of tool use in bottlenose dolphins. *Proceedings of the National Academy of Sciences, 102*(25), 8939–8943. doi:10.1073/pnas.0500232102

Lloyd, E. (2004). Kanzi, evolution, and language. *Biology & Philosophy, 19*(4), 577–588. doi:10.1007/sBIPH-004-0525-3

MacDonald, S. E., & Ritvo, S. (2016). Comparative cognition outside the laboratory. *Comparative Cognition & Behavior Reviews, 11*, 49–62. doi:10.3819/ccbr.2016.110003

Meyer, C. B. (2001). A case in case study methodology. *Field Methods, 13*(4), 329–352. doi:10.1177/1525822X0101300402

Miklósi, Á., Kubinyi, E., Topál, J., Gácsi, M., Virányi, Z., & Csányi, V. (2003). A simple reason for a big difference: Wolves do not look back at humans, but dogs do. *Current Biology, 13*(9), 763–766. doi:10.1016/S0960-9822(03)00263-X

Morgan, C. L. (1894). *An introduction to comparative psychology.* London, UK: Water Scott Ltd.

Nakata, K. (2007). Prey detection without successful capture affects spider's orb-web building behaviour. *Naturwissenschaften, 94*(10), 853–857. doi:10.1007/s00114-007-0264-9

Patterson, F. G., & Linden, E. (1981). *The education of Koko.* New York, NY: Holt, Rinehart and Winston.

Pepperberg, I. M. (1999). *The Alex studies: Cognitive and communicative abilities of grey parrots.* Cambridge, MA: Harvard University Press.

Pepperberg, I. M. (2001). Lessons from cognitive ethology: Animal models for ethological computing. *Learning, 35*(5), 5–12.

Pepperberg, I. M., & Funk, M. S. (1990). Object permanence in four species of psittacine birds : An African Grey parrot (*Psittacus erithacus*), an Illiger mini macaw (*Ara maracana*), a parakeet (*Melopsittacus undulatus*), and a cockatiel (*Nymph icus hollandicus*). *Animal Learning & Behavior, 18*(1), 97–108. doi:10.3758/BF03205244

Perdue, B. M., Talbot, C. F., Stone, A. M., & Beran, M. J. (2012). Putting the elephant back in the herd: Elephant relative quantity judgments match those of other species. *Animal Cognition, 15*(5), 955–961. doi:10.1007/s10071-012-0521-y

Pfungst, O., Stumpf, C. (Trans.), & Rahn, C. L. (Trans.). (1911). *Clever Hans (the horse of Mr. Von Osten): A contribution to experimental animal and human psychology.* New York, NY: Henry Holt.

Ringhofer, M., & Yamamoto, S. (2017). Domestic horses send signals to humans when they face with an unsolvable task. *Animal Cognition, 20*(3), 397–405. doi:10.1007/s10071-016-1056-4

Roberts, W. A. (2002). Are animals stuck in time? *Psychological Bulletin, 128*(3), 473–489. doi:10.1037/0033-2909.128.3.473

Rollin, B. E. (1986). Animal consciousness and scientific change. *New Ideas in Psychology, 4*, 141–152. doi:10.1016/0732-118X(86)90001-2

Rutishauser, R., & Moline, P. (2005). Evo-devo and the search for homology ("sameness") in biological systems. *Theory in Biosciences, 124*(2), 213–241. doi:10.1007/BF02814485

Scharff, C., & Petri, J. (2011). Evo-devo, deep homology and FoxP2: Implications for the evolution of speech and language. *Philosophical Transactions of the Royal Society of London. Series B, Biological Sciences, 366*(1574), 2124–2140. doi:10.1098/rstb.2011.0001

Shettleworth, S. J. (1975). Reinforcement and the organization of behavior in Golden Hamsters: Hunger, environment, and food reinforcement. *Journal of Experimental Psychology: Animal Behavior Processes, 104*, 56–87. doi:10.1037/0097-7403.1.1.56

Shettleworth, S. J. (2010). Clever animals and killjoy explanations in comparative psychology. *Trends in Cognitive Sciences, 14*(11), 477–481. doi:10.1016/j.tics.2010.07.002

Shubin, N., Tabin, C., & Carroll, S. B. (1997). Fossils, genes and the evolution of animal limbs. *Nature, 388*(6643), 639–648. doi:10.1038/41710

Shubin, N., Tabin, C., & Carroll, S. B. (2009). Deep homology and the origins of evolutionary novelty. *Nature, 457*(7231), 818–823. doi:10.1038/nature07891

Sorge, R. E., Martin, L. J., Isbester, K. A., Sotocinal, S. G., Rosen, S., Tuttle, A. H., ... Leger, P. (2014). Olfactory exposure to males, including men, causes stress and related analgesia in rodents. *Nature Methods, 11*(6), 629–632. doi:10.1038/nmeth.2935

Suen, H. K., & Lee, P. S. C. (1985). Effects of the use of percentage agreement on behavioral observation reliabilities: A reassessment. *Journal of Psychopathology and Behavioral Assessment, 7*(3), 221–234. doi:10.1007/BF00960754

Timberlake, W. (1997). An animal-centered, causal-system approach to the understanding and control of behavior. *Applied Animal Behaviour Science, 53*(1–2), 107–129. doi:10.1016/S0168-1591(96)01154-9

Trestman, M. (2015). Clever Hans, Alex the Parrot, and Kanzi: What can exceptional animal learning teach us about human cognitive evolution? *Biological Theory, 10*(1), 86–99. doi:10.1007/s13752-014-0199-2

Triana, E., & Pasnak, R. (1981). Object permanence in cats and dogs. *Learning & Behavior, 9*(1), 135–139. Retrieved from http://www.springerlink.com/index/F70857463613G101.pdf

Vonk, J., & Povinelli, D. J. (2006). Similarity and difference in the conceptual systems of primates: The unobservability hypothesis. In E. A. Wasserman & T. R. Zentall (Eds.), *Comparative cognition: Experimental explorations of animal intelligence* (pp. 363–387). New York, NY: Oxford University Press. Retrieved from http://books.google.com/books?hl=en&lr=&id=pf78r-F4FjIC&oi=fnd&pg=PA363&dq=Similarity+and+Difference+in+the+Conceptual+Systems+of+Primates:+The+Unobservability+Hypothesis&ots=C6yjRba893&sig=yyeQ1FA9K2ONXdKHlesPJWfkJK8

Watanabe, T. (2000). Web tuning of an orb-web spider, *Octonoba sybotides*, regulates prey-catching behaviour. *Proceedings of the Royal Society of London B: Biological Sciences, 267*(1443), 565–569. doi:10.1098/rspb.2000.1038

Wasserman, E. A. (1993). Comparative cognition: Beginning the second century of the study of animal intelligence. *Psychological Bulletin, 113*, 211–228. doi:10.1037/0033-2909.113.2.211

Yin, R. K. (2013). *Case study research: Design and methods* (5th ed.). Thousand Oaks, CA: SAGE Publications.

Consciousness in Animals

NOT ALL CONSCIOUSNESS IS THE SAME

When you see the word *consciousness*, what comes to mind? For most, it's like an on-off switch—either you're conscious or you're not. Whether it's from being knocked out by a head injury, being in a coma, asleep, or under anesthesia, or perhaps even finding yourself "zoned out" in a boring meeting, we frequently talk about being "not conscious." On the first day of my (ECW) *Exploring Animal Minds* seminar, I ask students to define consciousness as they understand it. Responses range from "being aware" to "responding in noninstinctual ways" to even "a word humans made up to feel special about themselves." We would be hard-pressed to find at least one scholar who would disagree.

For centuries, philosophers, neuroscientists, psychologists, and many others have attempted to define consciousness in humans. This is one reason for the wide range of answers to what consciousness is. Depending upon who you are, what your agenda is, and how you were trained, definitions for consciousness will vary. While there is little consensus, one major divide that guides this chapter's discussion of consciousness is the distinction between psychological versus phenomenal consciousness. Philosopher David Chalmers (1996) provides this helpful distinction:

Psychological	Phenomenal
Third person	First person
Behavior	Feeling
Objective	Subjective

For example, when someone is awake, as third-party observers, we can objectively observe awakeness by measuring brain waves. Thus, we might think of being awake, alert, attentive, or other responsive states as psychological consciousness because it can be measured or observed. On the other hand, when you stub your toe, a third-party observer cannot measure or access what you're subjectively feeling. This, we refer to as phenomenal consciousness.

To further distinguish Chalmer's psychological and phenomenal consciousness, try the following exercise. Slam your open hand down on a table. Yes, really! At a phenomenal consciousness level, you might feel sharp pain immediately, followed by a radiating or throbbing dull pain and some warmth, or absolutely nothing if you have nerve damage or congenital insensitivity to pain. No one has access to your subjective feeling but you. You are having a phenomenal conscious experience. On the other hand, a third-party observer who specializes in anatomy could tell you a great deal about how the sharp pain you felt was due to alpha-delta nerves sending a signal to your brain quickly, and the dull pain was due to c-fibers, which remind you of the injury until it heals. The anatomist could also tell you about the brain structures involved with you deciding to slam your hand on the table in the name of science in the first place—but we digress. The point is, even with all of this physical information about the body's pain pathway and nerve fibers, the anatomist could never tell you what it *felt* like for *you*.

Exploring consciousness in other species is even trickier for two reasons. First, there's the structure–function problem. While there's no guarantee that two people would experience an identical subjective feeling when they slam their hands on the table, there's a strong chance that their description would be similar due to similar anatomy. But even though we share much of the same anatomy with other species, there's still no guarantee that if a swordfish slammed its pectoral fin down on the table, it would experience the same feeling, or any feeling at all for that matter.

As philosopher Thomas Nagel famously put it in his thought experiment on bats, an organism is conscious "if and only if there is something that it is to *be* that organism—something it is like *for* the organism" (Nagel's emphasis, 1974, p. 436). While some, like philosopher Jeremy Bentham, would say basic pain and suffering are enough to qualify as Nagel's "something it is like," higher order behavioral correlates of consciousness (BCCs) like metacognition and self-awareness are more commonly studied by animal cognition researchers.

If you've ever had the experience of driving a car and arriving at your destination only to realize you were "zoned out," with no recollection of how you arrived there because your mind was elsewhere, you can see how an outward behavior can look complex and highly cognitive but can be carried out without conscious experience. Likewise, the rare condition of blindsight offers a similar, albeit permanent, human example of nonconscious experience. Humans have two visual systems. The first system is located deep within the ancestral brain and detects basic motion and shadows. The second system is the visual cortex, which is responsible for conscious visual experiences. Patients with blindsight have experienced serious damage to their visual cortex, perhaps from a car accident, so for all intents and purposes, they are blind. They will always report they cannot see anything. However, when they're asked to guess the direction an object moves on a computer screen or to walk down a cluttered hallway, they can do both successfully—much to their own surprise. This is because their eyes are sending basic visual information to the ancestral visual system. There's no subjective *feel* to their vision—they swear they cannot see anything—yet they respond in ways that appear to an outsider as if there is.

Though we might not like to admit it, there's a very real possibility that animals experience the world like a blindsight patient. The dog playing fetch in the backyard may be responding reflexively to an intense prey drive, and not experience the softness of the grass, the warmth of the sunshine, or anything subjective that goes along with it. Just like the blindsight patient experiences no conscious vision, animals could have the same situation for all their senses where they respond to their environment in ways that look similar to our own subjective, consciousness-driven behaviors, but they could really be feeling nothing. As Nagel put it, there may be nothing there is to be like an animal.

Researchers have tried to address this controversy by looking at the brain. Relying on the structure–function argument by analogy from Chapter 1, Historical Perspective on Animal Cognition, neuroscientists can ask people to report on their subjective experiences in the moment, determine their brains' corresponding neural activity, and then make inferences about what an animal might experience when the animal shows similar brain activity. Of course, correlation does not mean causation. I cannot conclude that an animal is conscious just because it has a brain structure that is associated with consciousness in humans. But it does lend strong support to conclusions made by animal cognition researchers. So strong, in fact, that neural structures associated with consciousness were the foundational evidence in the Cambridge Declaration on Consciousness, a 2012 collaboration between psychologists, neuroscientists, and other great thinkers. This two-page document concluded that it's very likely many animals, including birds, mammals, and even octopuses, are conscious.

One way that some scientists have gotten around the structure–function argument by analogy problem is to completely avoid the brain at all. Neuroscientist Giulio Tononi has been at the forefront of this movement. Just like mass or electrical charge, Tononi hypothesizes that consciousness might also be a universal property of all living things. He developed a mathematical formula that converts an animal's complexity into a numerical value he calls phi (Tononi, 2004). Simplified, phi is an organism's ability to integrate information. An animal like a worm, for example, can integrate some information, but not nearly as much as a mouse, and definitely much less than a dolphin or human. For over a decade, Tononi experimented with what he meant by "information" and how it would be converted into a mathematical equation. After a great deal of tinkering, Tononi inputted some real human brain activity data into his equation, and his formula seemed to hold up in some cases (Kim et al., 2018). With time, there's a strong possibility that his integrated information theory (IIT), or something very similar to it, will offer concrete answers to what consciousness is and how it can be measured.

Turning attention back to animals, if we define consciousness as a quantifiable value on a spectrum, it eliminates the need for consciousness to be an on–off switch. IIT would say that beetles, sloths, and humans are all conscious, just at different levels. This conceptualization allows consciousness to vary based on what a particular species needs for survival, which fits nicely within Darwin's framework of evolution by natural selection. It also opens the door for artificial intelligence to be considered conscious—a controversy that is far beyond the scope of our book, but exciting to ponder!

With this orientation to consciousness, let's delve into four BCCs. Each is relevant to how we view ourselves as distinct entities with unique perceptions, thoughts, and subjective feelings.

THROUGH THE LOOKING GLASS

The fact that it takes human children around 2 years to recognize themselves in a mirror is a testament to how much complex cognition is required to do it. At some point, however, all neurotypically developing children can look into a mirror and realize the image they see is "me" (Rochat, 2003). But what or who is "me"? To have a concept of "me" means to have a concept of self—that I am a distinct entity with thoughts, feelings, and desires. One who operates on the world as an independent being. This is different from self-consciousness, for example, which is like being mortified to go out in public with a bad haircut. As you'll see shortly, tests designed to assess self-awareness in animals have resulted in some surprisingly inconsistent findings. These inconsistencies have led researchers to reevaluate the methods used to measure self-awareness and what it means for a species to pass or fail those tests.

In 1970, Gordon Gallup placed a red mark on an anesthetized chimpanzee's forehead, then gave the animal a full-length mirror. The chimpanzee looked into the mirror and began to rub the mark off its own face, and thus, the mirror mark test for self-awareness was born. Most animals see their reflection as another animal and respond accordingly—usually with aggression or curiosity. Rather than treating the ape in the mirror as a threat, Gallup's chimpanzee appeared very confident that he was the guy in the mirror, and his face was dirty! Gallup concluded that the chimpanzee's behavior demonstrated a "rather advanced form of intellect ... [that] implies a concept of self" (Gallup, 1970, p. 87), in addition to consciousness and a mind (Gallup, 1982).

Though Gallup and others are convinced by his data, some critics outright reject the mirror mark test's ability to tell us anything about self-awareness (see Brandl, 2018). Despite these criticisms, the mirror mark test has been conducted for almost 50 years on more than a dozen different species. Before you start reading the findings, we invite you to pause and predict performance for monkeys, crows, and ants based on what you already know about these animals. When you're ready, move on, but we warn you the results might surprise you!

Among nonhuman primate apes, bonobos, orangutans (*Pongo pygmaeus*), and chimpanzees pass the mirror mark test. Gorillas' (*Gorilla gorilla*) performance is much less conclusive, leaving researchers puzzled as to why. One theory requires us to think about the society gorillas live in. For gorillas, prolonged eye contact is viewed as a threat (not the case in other ape species); therefore, they may perform inconsistently on the test because some might not look at their mirrored image long enough to realize they're looking at themselves.

Unlike the great apes, monkeys consistently fail the mirror mark test (see Anderson & Gallup, 2015 for review), a result dating all the way back to Gallup (1970) replicating his original chimpanzee study with rhesus macaques (pronounced REE-sus muh-KAK, *Macaca mulatta*). De Waal, Dindo, Freeman, and Hall (2005) have, however, shown that capuchin monkeys (*Sapajus apella*) do not treat the mirror image as a complete stranger, and with intensive training, some monkeys can be trained to respond to a mark on their faces when presented with their mirror image (e.g., Roma et al., 2007). The fact that they're being trained to respond, however, makes the results questionable.

Bottlenose dolphins (Reiss & Marino, 2001), killer whales (*Orcinus orca*), false killer whales (*Pseudorca crassidens*; Delfour & Marten, 2001), and Asian elephants (Plotnik, de Waal, & Reiss, 2006) all appear to recognize themselves. Horses spent time looking behind the mirror and engaging in behaviors that might seem promising (e.g., opening mouth, scraping cheek with mark against surface), but the authors concluded there was no clear evidence that they recognized themselves (Baragli, Demuru, Scopa, & Palegi, 2017). Horowitz (2017) opted to modify Gallup's task to be smell based rather than vision based to accommodate dogs' exceptional noses. She found that the dogs behaved differently toward a canister containing their own urine compared to a stranger's—leading her to conclude both the importance of the ecological validity of a task (i.e., matching the task to the species' natural history) and that dogs may actually have self-awareness. One of the only solitary (i.e., nonsocial) mammals tested, giant pandas (*Ailuropoda melanoleuca*), failed the test as evidenced by aggression toward their reflection (Ma et al., 2015), which indicates that self-awareness may be more important to social species than solitary species.

Among birds, there are also mixed results, possibly due to methodological differences. African grey parrots (Pepperberg et al., 1995) and New Caledonian crows (Medina, Taylor, Hunt, & Gray, 2011; *Corvus moneduloides*), which are two of the most likely candidates, have failed. Jackdaws (*Corvus monedula*), which are in the crow family, also have not passed (Soler, Pérez-Contreras, & Peralta-Sánchez, 2014), but two other crow cousins, magpies (Prior, Schwarz, & Güntürkün, 2008; *Pica pica*) and Clark's nutcrackers (Clary & Kelly, 2016; *Nucifraga columbiana*), have passed, which really leaves researchers scratching their heads.

As if that is not confusing enough, there are conflicting results for fish, too. At least one fish, the daffodil cichlid (*Neolamprologus pulcher*), failed, despite the authors predicting that they might pass since they are highly social (Hotta, Komiyama, & Kohda, 2018). Another highly social fish, the giant manta ray (*Manta birostris*), was tested because it has the largest brain size to body ratio of all fish species (Ari & Agostino, 2016). The manta rays behaved as if they might recognize themselves by performing repetitive movements in front of the mirror the same way you might watch yourself sway back and forth in a security camera's live video feed at an electronics store. In yet another twist, three different species of ants picked at small blue dots on their heads when they were presented with a mirror, but not without the mirror or when the color of the dot was the same color as their bodies (Cammaerts & Cammaerts, 2015). The surprised researchers could only speculate about what their findings meant, but they were pretty sure of the fact that ants can recognize at least their bodies as their own. Whether or not ants have true self-awareness that extends beyond body ownership to an idea of being a distinct individual with an identity is less clear.

Unexpected results have led many researchers to criticize the mirror test's ability to teach us anything about which animals have self-awareness and which ones do not (e.g., Brandl, 2018). The challenge of the mirror mark test also raises the issue of anthropomorphism, the attribution of humanlike traits to animals. It is entirely possible that species who fail actually do have a sense of self, but they may not care about a mark on their bodies in the same way that humans do. Truth be told, some humans would not care either! In this way, the mirror mark test may be more of an indicator of how important grooming is to a species, a simpler conclusion that Morgan's Canon would tell us to investigate first prior to jumping to the complex state of self-awareness.

Other criticisms of the test come from those who find it strange that there are such great inconsistencies in sample size and overall species' performance. Impressively, Ma et al.'s (2015) giant panda study had 34 subjects, while Ari and Agostino's (2016) manta ray study only had two. While power studies with small sample sizes can be incredibly helpful to illustrate what is possible for a species, it is hard to make strong conclusions when the sample size is low *and* there is inconsistency. For example, only three of the five magpies in Prior et al.'s (2008) study passed. Imagine if we equated this to humans; it would mean 40% of the adult population could not recognize themselves in a mirror and presumably do not have a self-concept, which simply does not seem reliable as a result.

Despite the criticisms, one thing is certain: the mirror test still remains a gold standard used by researchers who study animal consciousness. Perhaps, rather than fixating on the problems with the method, we might instead turn our attention to what it would mean to us if it is true that ants have a self-concept, for example. Would that make you think twice about laying down ant traps in your home or flicking them off your picnic bench?

AWARENESS OF OTHERS

One of the most fascinating milestones in child development is when kids figure out how to tell a lie. To effectively pull off a lie, a child must be able to separate what they know about the world from what another person knows and does not know about it. Then, they need to create a false reality and present it persuasively to the other person, all the while keeping both the real and fake realities straight in their own heads. Broken down like this, we hope it's clear that lying is really no easy feat!

Given the cognitive complexity of lying, it should come as no surprise that in addition to practice, children get better at lying as their brains mature and they become better able to separate themselves from others. The age at which children begin to successfully lie is around the time they begin to refine their theory of mind

(Premack & Woodruff, 1978). Theory of mind is the ability to view your own knowledge and experiences as being distinctly different from someone else's, which is fundamental to lying. For example, when Jane lies to her grandpa about how much money she gave herself from the Monopoly game bank, she can get away with it because her theory of mind allows her to realize that grandpa would never know she took extra since he was in the bathroom at the time.

Another way of thinking about theory of mind is perspective taking. Sometimes we put ourselves in the perceptual shoes of others, such as not wearing a large hat to the movie theater so those behind us can see. Other times, we put ourselves in the cognitive shoes of others, such as realizing the neighbor has no clue it was you and not your dog who stepped in his flower bed. Both examples require taking someone else's perspective, which gives us the ability to engage in complex social interactions such as empathy and deception.

Theory of mind may be a hallmark of humanity, "central to everything that makes us human" (Krupenye, Kano, Hirata, Call, & Tomasello, 2016, p. 110), but how and why it evolved has scientists stumped. Some say our perspective-taking ability helped our human ancestors manage social relationships in our increasingly larger group sizes. Others view theory of mind as somewhat of a random by-product of our large brain size. Although there is much debate, there is some evidence leaning toward the latter (see Devaine et al., 2017).

Just as there is no consensus on why we can put ourselves in others' shoes, we also do not know whether the ability is uniquely human. When Premack and Woodruff (1978) first published their idea on theory of mind, they argued chimpanzees had a theory of mind that was similar to a human's. Their classic goal-understanding experiment involved showing a 14-year-old chimpanzee named Sarah a series of video clips of a human experiencing problems such as being locked in a cage or reaching for but unable to grab some bananas. After watching the clips, Sarah was presented with two photographs, one of which was the solution to the problem (e.g., a key to the cage or a stick to reach the bananas). When faced with the choice between the two photographs, Sarah was more likely to pick the logical solution to the problem. According to Premack and Woodruff, this showed that Sarah could put herself in the shoes of the person in the video in order to understand their predicament, then respond accordingly.

More recently, others have recorded chimpanzees' behavior toward situations in which the experimenter was either unwilling or unable to give them food. In one study, trials included a clumsy experimenter who "accidentally" dropped a grape that they were trying to hand to the animal, a blocked experimenter who tried to hand a grape to the subject but was unable to squeeze it through a hole to pass it to them, a rude experimenter who teased the subject by starting to pass the grape to them then pulled it back, and a refusing experimenter who sat a grape in front of them and then stared at the chimpanzee (Call, Hare, Carpenter, & Tomasello, 2004). As expected, the subjects behaved in ways that suggested they could tell when the experimenter had no intention of giving them a grape; they made more of a ruckus, banged on the apparatus, and gave up and left the testing area sooner.

Animal theory of mind studies do not always involve humans calling the shots. Chimpanzees have also been documented deceiving each other and sometimes humans. Kirkpatrick (2007) describes a subordinate male "hiding" his erection from a dominant male to avoid a conflict and a female sitting on food that she found until a thieving chimpanzee had left the area. One famous chimpanzee, Santino, displayed strong indicators of deception toward human visitors at Furuvik Zoo in Sweden (Osvath & Karvonen, 2012). Guests reported Satino threw rocks and other projectiles at them. When the researchers observed him, they discovered that not only had he been hiding objects under self-made hay piles and behind large rocks in his enclosure, but he would also approach visitors calmly, get as close as possible, then throw objects before the visitors had time to get away. This is particularly fascinating because chimpanzees do throw projectiles, but to approach the zoo visitors first in a nonaggressive way as if there was a goal to be initially nonthreatening suggests a highly sophisticated deceptive attack.

Given the thousands of years that dogs have spent with humans, you'd expect them to be masters of reading our minds. Most dog owners have probably felt like their dog "understands them" or "can tell" when they are sad. However, despite being very highly tuned to our behavior, they do not seem to be able to put themselves in our shoes (Bräuer, 2014). They do seem able to know when we are trying to communicate something to them, which Bräuer notes other species have great difficulty with. This has made some scientists conclude that dogs are great at reading the subtleties of our behavior, just not our minds (Udell & Wynne, 2011). Nonetheless, at least one study has provided compelling results for theory of mind. The authors found that dogs were more likely to steal "forbidden" food when an experimenter's back was turned, eyes were closed, or eyes were covered by a blindfold (Maginnity & Grace, 2014), possibly indicating visual perspective taking.

Imagine you have something very valuable. What would you do if a stranger caught you in the act of hiding that treasure? Likely, you might wait until the stranger left and then hide it somewhere else so it does not get stolen. This is exactly what corvids seem to do, too. In one experiment, ravens (*Corvus corax*) buried food faster and did a higher quality job of hiding it when there was a chance a thieving raven could be watching them through a peephole compared to when the peephole was shut (Bugnyar, Reber, & Buckner, 2016). Similarly, western scrub jays (*Aphelocoma californica*) reburied their food if another animal was present (Dally, Emery, & Clayton, 2006). What's even *more* interesting is that jays that had previously been thieves themselves reburied their food more often than nonthieves—suggesting they knew what their rivals were capable of given their own previous mischief (Dally et al., 2010).

Despite these fascinating findings, there are still those who are not convinced and argue that the animals are responding based on simpler behavioral rules, rather than an ability to actually project themselves into the minds of others (e.g., Povinelli & Vonk, 2004; van der Vaart, Verbrugge, & Hemelrijk, 2012). Because we cannot ask animals to explain their reasoning, we are at a bit of a standstill as to whether there is higher order complex cognition happening. The best we can do is be very creative and careful when designing perceptual and cognitive theory of mind tasks and rule out simpler alternative explanations.

Another challenge that researchers must reconcile is the fact that no one species has been able to pass all of the different theory of mind tasks out there (Horowitz, 2011). What might this mean? Further, many of the tasks require the animal to engage with a human experimenter. What if a particular species does not pass the test because they do not view humans as appropriate social partners? Finally, up until this point, we have been discussing theory of mind as if either you have it or you do not. What if the ability to take the perspective of others is on a spectrum the way some speculate about consciousness in general? This is also a possibility that some have explored (see Horowitz, 2011).

REFLECTING ON WHAT YOU KNOW

What did you have for lunch 3 days ago? Now, on a scale from 1 to 10, 10 being 100% sure, how confident are you about your response? In order to answer this second question, you must engage in metacognition—thinking about your own thinking or reflecting on your own knowledge. It requires more than remembering by adding an additional layer of confidence or sureness. To proclaim, "I am pretty sure I had panang curry for lunch, but I'm not positive, so I'd rate it a 9," is to have activated metacognitive processes.

Being able to self-reflect and think about what one knows allows an organism to more effectively respond to their environment. For example, if a friend asks a few weeks later how to get to the restaurant that served the panang curry, through the process of metacognition, I can reflect upon my knowledge and confidence in giving directions and respond accordingly. As I reflect, areas in my brain related to my concept of self are also activated, showing just how interconnected these BCCs really are (Metcalfe & Schwartz, 2016).

Since we cannot ask animals to verbally reflect, metacognition researchers have come up with ways to tell if an animal can think about its own thinking. One way is to give them the opportunity to ask for more information. For the restaurant directions example, pushing a button might let me see the restaurant's address. For animals in laboratory settings, this might mean having a stimulus presented again or receiving some new information prior to making their official response (Kirk, McMillan, & Roberts, 2014). As you might have predicted, subjects are more likely to ask for "hints" when the questions are more difficult, which is a behavioral indicator of the animal's metacognitive processing.

A second commonly used metacognition task is the uncertain option task (e.g., Smith, Shields, Schull, & Washburn, 1997). Smith et al. tested one dolphin on an auditory discrimination task with an "opt out" option. The target stimulus was a 2,100 Hz high-pitched tone. If the dolphin heard this tone, he needed to press a left paddle; if he heard any other tone, the right paddle was the correct choice. As the trials got more and more difficult, Smith and colleagues noticed that the dolphin was more likely to make use of a third paddle, which allowed him to opt out of the trial and replace it with an easier one. For example, a 2,150 Hz tone is very close to the 2,100 Hz target. The dolphin who had trouble distinguishing the two might "opt out" of responding—thereby asking for a new test tone to replace the 2,150 Hz one—rather than guessing and potentially giving the incorrect answer.

Finally, sometimes metacognition researchers get animals to report on their confidence using wagering. Here, the subject must reflect on its own knowledge following its answer on a memory task and offer either a high- or low-risk bet. For a low-risk bet, the animal might receive one guaranteed token (i.e., object that can be turned in for food reward later), but for the high-risk bet, they either receive three tokens for a correct response or lose three for an incorrect response. Based on how confident the animal is in their performance from trial to trial, the animal should respond so as to maximize its tokens. This task is essentially the Final Jeopardy round, where it would be silly for a contestant to risk it all when they are not totally confident about their response.

Unlike many Jeopardy! contestants, most nonhuman apes seem to "know what they do not know" (see Call, 2010; Call & Tomasello, 2008). In one comparative experiment, children and chimpanzees observed one of two cups get filled with a food reward; then, they had the opportunity to choose which cup they would like to receive (Nelder, Collier-Baker, & Nielsen, 2015). When the cup was filled out of view (meaning the subject did not know which cup held food), both the children and chimpanzees were more likely to choose a third, guaranteed small reward cup. The authors concluded that the chimpanzees and children were able to reflect upon their knowledge of the two cups and respond with the third when their chance of being correct went down.

Monkeys also seem capable of metacognition. They'll ask for more information and use an uncertainty response (Beran, Perdue, Church, & Smith, 2016; Beran & Smith, 2012; Rosati & Santos, 2016). They will also make confidence judgments on memory tasks that reflect what appears to be perceived confidence (Morgan, Kornell, Kornblum, & Terrace, 2014). Nonetheless, while these findings point toward metacognitive abilities in monkeys, it's important to note that there have been negative findings as well (e.g., Paukner, Anderson, & Fujita, 2006).

It may not bring defeated Jeopardy! contestants much comfort to know that the common brown rat (*Rattus norvegicus*) has shown a remarkable ability to pass metacognition tests. In one experiment, rats learned to pull a lever to receive a hint as to which end of a maze contained food. On top of that, when the reward location was always on the same side, the rats stopped asking for hints, presumably because they figured out the pattern and were more confident in their choice (Kirk et al., 2014). In a similar task, Templer, Lee, and Preston (2017) exposed rats to an odor, then there was a time delay, and then they had to select the odor they had been exposed to from a set of four odor options. Here, just like with great apes, the rats were more likely to opt out of taking the odor memory test as the difficulty increased. Kudos to Templer et al. for designing an ecologically relevant task that played off rats' terrific sense of smell rather than using, for example, a visual test (rats have poor eyesight).

Interestingly, there has been little published on canine metacognition; however, another ecologically relevant study played off dogs' strong social bond with humans and used an actual human informant instead of a button or lever (McMahon, Macpherson, & Roberts, 2010). With this setup, dogs were more likely to go to the human who was pointing to the location of food as opposed to a human who simply stood nearby. The authors were hesitant to conclude that they had clear evidence for metacognition, but they did not rule it out as a possible explanation for the dogs' behavior.

Though some of the findings are inconsistent, some bird species seem to reflect on their knowledge as well (Iwasaki, Watanabe, & Fujita, 2018). Western scrub jays ask for more information when they are unsure of where a human placed food (Watanabe & Clayton, 2016). One unsuspecting bird that has also passed is the pigeon (*Columba livia*). Iwasaki et al.'s (2018) subjects had to memorize lists of three familiar or unfamiliar images (e.g., clip art shapes and pictures). The pigeons were more likely to ask for a hint in advance of having to memorize unfamiliar lists compared to familiar lists. This led the authors to conclude the pigeons were capable of understanding they did not know the unfamiliar lists and respond in a way that would help them perform more accurately later on by requesting a hint.

We saved the best for last ... even honeybees (*Apis mellifera*) will opt out of difficult trials if they are given the opportunity (Perry & Barron, 2013). Though it could be an act of species bias, some researchers point out that if even honeybees can succeed, then any animal with a nervous system has the basic decision-making brain structures needed to opt out or seek more information. The fact that so many nontraditional species like bees and pigeons can pass metacognition tasks but not self-awareness or theory of mind tasks does seem fishy, after all (e.g., Insabato, Pannunzi, & Deco, 2016). Maybe metacognition tasks just tap into an evolutionarily adaptive and widespread decision-making system—not some complex cognitive process grounded in consciousness.

Despite criticisms, the ability to reflect upon one's own knowledge does appear to have some special quality to it. New directions have attempted to link outward metacognitive behavioral responses to real-time neural activity. For example, Miyamoto et al. (2017) found that when a particular area of the prefrontal cortex was impaired, monkeys' ability to remember information was fine, but their ability to judge their confidence for those memories was disrupted. This is an exciting finding because it clearly shows that there could be more to animals' opting out or hint seeking than critics think.

EMOTIONS

Sometimes research takes you to the field, sometimes to the laboratory, and other times to haunted houses. Patient S.M. (so named to protect her identity) suffers from an extremely rare genetic condition called Urbach–Wiethe disease, which involves severe damage to the brain's amygdala. As a result, S.M. cannot experience the emotion of fear. To test this, researchers tried to call her bluff by taking her to the Waverly Hills Sanatorium—dubbed one of the most terrifying haunted houses in the world. S.M. encouraged the group of terrified scientists to keep up with her as she boldly rounded corners and tried to strike up casual conversations with the monsters that jumped out at her (Feinstein, Adolphs, Damasio, & Tranel, 2010). At no point did S.M. report any emotions besides excitement and curiosity. In fact, besides the researchers, the only other individual who reported being scared during the experience was one of the haunted house actors whom S.M. poked in the head because she was curious to know what the monster felt like.

Fear is just one of the many subjective experiences that contribute to the richness of what it means to be human. We call these internal states emotions, or short-term moments of feeling in response to internal or external stimuli. Some argue that emotions are critical to humans' ability to navigate the world as decision-making beings (e.g., Damasio, 1995), which begs the question, do other species experience the subjective feeling of emotions too? Some would say no, but anyone who has spent a great deal of time with an animal will confidently tell you we are not the only ones who experience a range of positive and negative states (Walker, McGrath, Handel, Waran, & Phillips, 2014). The challenge for animal consciousness researchers is to come up with ways to test for emotions in animals, which is a feat that is easier said than done.

In what is considered to be the first systematic study of emotions in animals, Darwin's (1872) book *Expression of the Emotions in Man and Animals* outlined emotional expressions as adaptations for survival. For example, scrunching the nose in disgust would close off the nostrils and keep the individual from breathing in potentially harmful air (Chapman, Kim, Susskind, & Anderson, 2009). In addition to serving a survival function, Darwin used species ranging from cows to chameleons to show that outward expressions like nose scrunching are also important to signal information to social partners about their internal state, or emotion.

Darwin presented many interesting behavioral parallels between humans and animals with respect to emotions. For example, an enraged human might sneer, curling their lips back. Darwin likened this to an enraged animal baring its canine teeth as a threat. While present-day humans do not attack with their teeth, ancestral aspects of this useful emotional expression are still present and help to communicate the internal state of a very angry human (Hess & Thibault, 2009). Presumably, being able to express on the outside what is going on inside allows for quick communication such as, "I feel grumpy inside; leave me alone," or "I feel happy inside; I want to play." As you might imagine, there's survival and social value to emotional expression, regardless of species, which is why it seems logical for other animals to experience them.

Emotions allow us to communicate our internal states to others, but they have value beyond this as well. Even when there's no social partner present to communicate with, most would agree that emotions are still important. The feelings evoked from us when we hear our favorite song, ride a rollercoaster, or even attend a funeral enrich the constantly playing movie philosopher David Chalmers (1996) used to describe consciousness. One might go so far as to argue that without emotions, there would be no "something there is to be like" a human (Nagel, 1974). This is why some have argued that emotions represent a clear indicator of consciousness (Cabanac, 1999).

Most theories of animal emotions are based on careful naturalistic observation studies of behavior in particular contexts, be it in the field or in a laboratory setting. This might come in the form of tail flicks, body posture, elevated heart rate, scratching, ear position, repetitiveness, or head bobs. For example, perhaps a subordinate individual flicks its tail more when a dominant animal is released into the enclosure. By being informed about the species' natural history and when tail flicking tends to occur under other circumstances (such as when predators may be near), researchers might conclude that the animal is experiencing anxiousness when it flicks its tail.

Some of the most compelling evidence for basic emotions, like anxiousness, comes from observing how behaviors change when an animal is under the effect of drugs. For example, we might test the prediction that tail flicking indicates anxiousness by giving the animal a dose of an antianxiety drug and observing its behavior when it hears the dominant animal approaching the enclosure. If the animal being observed continues to explore the area and otherwise goes about its business with no tail flicking, it would provide additional evidence that the tail flicking really does correlate with an internal state of anxiousness. As an added control, the researchers could give the animal an anxiety-increasing drug and see if tail flicking is higher than normal when the dominant animal approaches the enclosure.

In addition to observing the effects of drugs, the cognitive bias task has also gained popularity (e.g., Paul, Harding, & Mendl, 2005). Humans who are depressed or anxious are more likely to make negative judgments about ambiguous situations. For example, if your current emotion is anxiousness or sadness and your significant other is not home from work at their usual time, you would be more likely to think something terrible happened to them. If you are in a positive internal state, you might think they've stopped at your favorite restaurant to pick up a romantic dinner for two. The same has been observed in a variety of species. Essentially, by exposing an animal to a repeated negative or positive experience, a future "pessimistic" or "optimistic" outlook on its environment can be induced. Now, when an ambiguous object is presented, just like with the late significant other, the pessimistic animal treats the object as aversive, whereas the optimistic animal wants to explore it.

Some believe emotions evolved after amphibians, making subjective internal states unique to reptiles, birds, and mammals (Cabanac, 1999), but recent research has also found fish and invertebrates may experience emotions as well. Thinking from an animal welfare perspective, given the number of animals in human care as pets, entertainment, working animals, research subjects, and livestock, finding out that animals experience emotions would be helpful to improving how we treat them. For example, in a cognitive bias study with rats, Harding, Paul, and Mendl (2004) found that rats repeatedly housed in an unpredictable (e.g., damp, tilted, undesirable) cage were less likely to explore their surroundings or take a chance at being wrong in a simple tone discrimination task—something most rats would normally excel at. Though this study focused on rats in a laboratory, the connection of the findings to other animals living in human care is clear.

Nonhuman primates (Bateson & Nettle, 2015; Maestripieri, Schino, Aureli, & Troisi, 1992), dogs (Mendl et al., 2010), starlings (*Sturnus vulgaris*; Bateson & Matheson, 2007), sheep (*Ovis aries*; Doyle, Fisher, Hinch, Boissy, & Lee, 2010; Reefmann, Wechsler, & Gygax, 2009), horses (Henry, Fureix, Rowberry, Bateson, & Hausberger, 2017), honeybees (Bateson, Desire, Gartside, & Wright, 2011), and even fruit flies (Deakin, Mendl, Browne, Paul, & Hodge, 2018) have all been shown to exhibit pessimism, negative affect, or low mood (terms vary depending upon the author). If this is true, it suggests that a wide range of animals are capable of negative emotional states, not just mammals or vertebrates.

It's one thing to refer to a negative state as pessimism, negative affect, or low mood, but it's another to evoke the term *learned helplessness*, which is commonly attributed to clinically depressed humans. Staring at the same fruit fly—or at least what I (ECW) think is the same one—day in and day out resting on my kitchen window, I have caught myself wondering if it eventually gives up on ever finding a way out of my home. First described using dogs (Seligman & Maier, 1967), learned helplessness is the resolve that nothing someone does will make their situation better, so they might as well give up. The idea that this common symptom of clinical depression was first understood by testing dogs may seem strange. On the other hand, it appears I was not too far off in my musings about the fruit fly on my window. It turns out fruit flies that have control over their environment and can turn off a mild electric shock by walking around learn to spend a lot more time walking, whereas fruit flies that receive random shocks (akin to the inescapable bad things over which we have no control in life) essentially give up and stop walking around (Batsching, Wolf, & Heisenberg, 2016).

Insects are not the only invertebrates to demonstrate emotional distress states. In one experiment, after being defeated by a more aggressive crayfish (*Procambarus clarkia*), subjects spent more time in a dark area compared to a well-lit area, a measure of animal anxiety (Bacqué-Cazenave, Cattaert, Delbecque, & Fossat, 2017). Interestingly, when the loser crayfish were injected with an antianxiety drug, they spent more time in the well-lit area. The authors even framed their results around the idea that crayfish may have a concept of defeat, which induces an anxiety-like emotional state. You can read more about the surprising world of invertebrate emotion research in Perry and Baciadonna's (2017) review.

Imagine if you made a career out of tickling rats and making them laugh. That's just what researcher Jaak Panksepp did for more than a decade. Panksepp found that when he tickled young rats all over their bodies, they would emit ultrasonic vocalizations and chase his hand around, as if to say, "More please!" Since play is considered to be a luxury, a behavior that is only observed when an animal has all its survival needs met, there is strong reason to believe that this "evolutionary antecedent" to laughter that Panksepp and Burgdorf (2003) uncovered is a real example of a positive emotional state in an animal.

Dogs (Andics et al., 2016), iguanas (*Iguana iguana*; Cabanac, 1999), and cows (*Bos taurus*; Proctor & Carder, 2015) also appear to experience positive affect or emotion. Emotions in another large mammal, the bottlenose dolphin, has been studied for welfare reasons due to their close and regular interaction with people and other dolphins when they are in human care. In two different studies, Clegg, Rödel, Boivin, and Delfour (2018) and Clegg, Rödel, and Delfour (2017) concluded that captive dolphins get excited when they expect to interact with humans and that dolphins that are more social with one another perform in more optimistic ways later on when they do cognitive bias tests.

Finally, just as honeybees can experience negative affect, they also appear to experience positive cognitive bias when they are unexpectedly given a drop of sugar water, the bee equivalent of your day looking brighter after randomly finding $10 on the sidewalk (Perry, Baciadonna, & Chittka, 2016). Though some advocate for serious caution and reframing of insect emotions as "motivational states" (e.g., Baracchi, Lihoreau, & Giurfa, 2017), those who conduct insect emotion research are highly convinced that their findings provide insights into the evolutionary building blocks of emotions.

The idea that there are biologically hardwired basic emotions throughout the animal kingdom is controversial, and for good reason (Bliss-Moreau, 2017). Just because a dog can move its face into a smile and just because humans smile when they are happy, it does not mean a dog must be happy when it smiles, too. Despite Darwin's use of similar expressions like smiling as evidence for common basic emotions across species, critics argue against such analogical leaps. There could be no actual feeling associated with the expressions that humans recognize as familiar, or if there is feeling, who's to say that it would be expressed in a way that people would recognize (Bliss-Moreau, 2017)? Without being able to obtain verbal self-report from animals, we run into the same challenge that plagues other areas of animal cognition.

To try to address the self-report problem, researchers interested in studying emotions in animals have tried to correlate behaviors with neural activity. Based on common neural circuitry, Jaak Panksepp (2005) identified fear, rage, seeking, lust, care, panic, and play as basic affective states (i.e., emotions) that are present in differing degrees based on each species' natural history. While Panksepp was largely referring to mammals, some think all vertebrates and some invertebrates might have this common neural circuitry for basic emotions (e.g., LeDoux, 2012). Panksepp was certain that without sacrificing methodological rigor, scientists could continue to study animal emotions, and the same could be said for all of the BCCs discussed in this chapter.

ANIMAL SPOTLIGHT: HAPPY THE ELEPHANT

As the saying goes, an elephant never forgets. Thus, many people are partial to claiming extraordinary memory when they think about these large mammals. It's true that field studies of wild elephants have shown that they're able to remember their way back to food and water resources hundreds of kilometers away. But neuroanatomical findings indicate that the brains of elephants would likely put them on par with humans with respect to their long-term memory capacity (Patzke et al., 2014).

Another similarity that elephants appear to share with humans is their ability to recognize themselves in a mirror. As we have seen, mirror self-recognition is connected to consciousness by way of self-awareness, or having a distinct concept of self as a separate entity in the world. In this "Animal Spotlight," we will share the story of Happy, a famous Asian elephant who made headlines with her mirror behavior.

Happy the elephant is presumed to have been born in 1971 in Thailand. In a time when it was still acceptable to remove animals from the wild for zoos, she, along with six other calves (all named for each of Snow White's seven small friends), was captured. In 1977, Happy and Grumpy arrived at the Bronx Zoo in New York City.

Elephants are well known for empathetic behavior (e.g., Byrne, Bates, & Moss, 2009), and given the established connection between self-awareness and empathy (e.g., Gallup & Platek, 2002; Plotnik, Lair, Suphachoksahakun, & de Waal, 2011), animal behavior researchers Plotnik et al. (2006) hypothesized that elephants would be a good candidate to test for self-awareness with the mirror mark test. Until that time, only dolphins, humans, and great apes had passed the test, but all four groups had also been well documented engaging in empathy-like behavior. Thus, Plotnik et al. set out to test Happy and the two other elephants living at the Bronx Zoo, Patty and Maxine.

The study consisted of one baseline observation phase, three carefully designed control phases, and a fifth experimental phase. In the baseline phase, Plotnik used an ethogram to record the elephants' baseline activity for certain target behaviors, like touching the head with the trunk. In the first two control phases, either the mirror was present but covered or it was visible. In the third control phase, an experimenter pretended to mark the elephants' foreheads with a large X and put them out in front of the covered mirror. In the experimental phase, the familiar mirror mark test was conducted; a real X was placed on the elephants' foreheads, and the cover over the mirror was removed. The number of target behaviors was able to be compared to the number of target behaviors during the baseline and control phases.

In the third control phase, the elephants engaged in rare behaviors such as reflection testing by moving their trunks and bodies in repeated, rhythmic ways in front of the mirror. They also displayed self-investigative behaviors such as moving their trunk to their mouth or ears in front of the mirror. These self-investigative behaviors gave the researchers confidence that all three elephants would pass the mirror mark test in the experimental phase.

Happy was the only elephant to pass the mirror mark test, though Maxine and Patty continued to engage in self-investigative behaviors using the mirror. Video recordings showed that once Happy had been marked, she immediately walked to the mirror where she stood for 10 seconds, then walked away. She returned 7 minutes later to test her reflection, then left again. Out of view of the mirror, she began touching the mark on her forehead, then returned to the mirror to touch and investigate the mark some more. Interestingly, when they tested her two more times on different days, including 2 months later, she did not respond to the mark on her forehead. Plotnik et al. (2006) suspect this could be due to the fact that elephants, unlike primates, do not tend to groom specific focal areas. It could be that once Happy realized the mark was not a danger to her, she was no longer concerned by it being on her body.

Happy and other animals in human care who have shown advanced cognitive abilities have launched a new branch of zoo research, that of welfare. Zoo welfare researchers aim to determine how to best understand and provide for an animal's needs (e.g., Melfi, 2009). Some of the challenge involved in this field has been moving away from anthropocentric tendencies. The idea that more space is always better, for example, is common; however, studies have shown that diversity in types of space (indoor, outdoor, shaded, sunny) and availability to choose when to go where is likely equally or potentially more important than actual square footage in some species (Fraser, 2009; Hill & Broom, 2009; Kurtycz, Wagner, & Ross, 2014; Tan et al., 2013).

Much of the focus on welfare has been on providing cognitive enrichment for animals. Largely credited to Hal Markowitz, enrichment can be anything which provides stimulation in some form—be it visual, tactile, social, olfactory, auditory, cognitive, or any other category (Markowitz, 1982; Markowitz & Stevens, 1978; Shepherdson, Mellen, & Hutchins, 1988). Current standards for accreditation by the Association of Zoos and Aquariums (AZA) require a formal enrichment plan be in place for every animal and that it is maintained by a staff member and appropriate records are kept (Accreditation Standards and Related Policies, 2017).

The week after Halloween is a great time to see enrichment in action at the zoo. Have you ever wondered what to do with your pumpkins? Bet you cannot guess what happens when an elephant gets a pumpkin to play with (actually, bet you can—and there are some great online videos). You know who loves unflavored gelatin?

Sea lions. Ever seen a chimpanzee use a tablet? These are all examples of daily enrichment provided to animals in human care at zoos. Sometimes enrichment can be elaborate—like the preceding cases. Sometimes it can be as little as a change in scenery. Many zoos and aquariums have gates or barriers that allow areas to be divided or groups of animals to be split. Rotating which area an animal is in and which other animals it is with (within social constraints and physical needs, of course) is, in many cases, enough to break up the day and increase activity levels (Tarou & Bashaw, 2007). Even in reptiles, enriched environments—ones with a variety of climbing, hiding, and basking spots—produce measurable difference in both behavior and health (Almli & Burghardt, 2006; Case, Lewbart, & Doerr, 2005). Training programs and research studies are also excellent cognitive enrichment—when positive reinforcement is used, a training session is more like a game of 20 questions without the questions. The trainer must communicate their request, and the animal must try to figure it out. Likewise, being exposed to a mirror for the first time, for example, is a great way to keep animals' minds stimulated and teach us something about the species.

Feeding enrichment is one of the more common and exciting ways of enriching animals in human care. Diets and treats can be scattered within trees or grass to create an opportunity for foraging, or they can be placed into puzzle balls for animals to extract. Even a program in which the public is allowed to feed an animal—under closely monitored circumstances—can be enriching. Not for the faint of heart, carcass feeding is also used as enrichment for predators like big cats or even lizards like the Komodo dragon. One of the first enrichment devices developed by Hal Markowitz involved stringing meatballs on a type of zip line and allowing carnivores to stalk and capture the moving prey (McPhee, 2002; Mellen, Markowitz, & Stevents, 1981).

Lastly, animals, like humans, love their toys. But dealing with a child who puts everything in his mouth is quite different than dealing with a bear or a penguin who does. Zoo toys tend to be objects—balls, drums, blocks—of hard plastic that can be carried around or "mouthed." Strips of robust tubing, foam, or vinyl are attached to things, and climbing ropes, hammocks, and ladders are popular for many species. Next time you're at the zoo, look around to see what types of creative enrichment are in the animals' enclosures, keeping in mind that it may even be part of a valuable research study like the one Happy participated in!

HUMAN APPLICATION: THEORY OF MIND IN CHILD DEVELOPMENT

As mentioned earlier, the point at which a child can effectively lie is a milestone moment in development. To lie, the child must manage both the current reality and a fabricated reality—and the former must be realistic enough to be accepted as truth. Little Regina might have accidentally broken the basement window while she was playing in the backyard, but if she can whip up some tears and say that the neighborhood bully did it, she can avoid punishment. Being able to take on the perspective of another is an extremely useful ability for any social species, both for antisocial and prosocial reasons. In this case, not only will Regina avoid being punished by her mother, but at a social level, lying about the incident allows her to maintain her positive relationship with her mother. Imagine the look on Regina's face if her mother found out she lied and responded with "Thank you so much for lying to protect our precious mother–daughter relationship" rather than punishing her. Try out that approach sometime, and let us know how it goes!

While that last consideration is silly, the underlying point is not: Theory of mind allows humans to have positive social interactions. Whether it is two romantic partners thinking about how their words might impact one another before speaking, a caregiver putting herself in the shoes of her child to better connect with them, or even a human anthropomorphizing to predict which toy their cat would like most, being able to take the perspective of others gives us a significant advantage when it comes to social interactions. By being able to predict others' behavior and mental processes, we can interact more effectively.

We're actually not born with theory of mind. While infants will look where others look and look where others point by about 1 year of age (Leung & Rheingold, 1981), it is not until they start to develop their concept of self that any of the processes related to theory of mind begin to unfold. This makes sense; in order for me to understand that I am distinctly different from others, I need to have a concept of "I" first. By about 2 years old, 65% of children are able to pass Gallup's (1970) classic mirror mark test (Amsterdam, 1972). Eventually, all neurotypically developing children pass.

Famed child psychologist Jean Piaget (1896–1980) spent a great deal of time studying his own children in order to create his four-stage theory of child cognitive development, which is still taught in developmental psychology courses today (Piaget, 1953). According to Piaget, toddlers would treat their reflection as another child until at least the end of his first stage, the sensorimotor stage, and for most children, into his second stage, the preoperational stage. It is in the preoperational stage that things start to get interesting.

The preoperational child who passes the mirror test still visibly struggles with separating themselves from others, a phase called *egocentrism*. In this phase, the child may hear "You're in the way!" over and over because they do things like stand in front of the television, blocking everyone's view. It's not done maliciously—*they* can see the television, so they assume *everyone* can see the television. With an increasing number of social interactions such as these, in addition to maturing brain development, the child will begin to understand how to separate herself from others and that she should check others' perspective on the television as well.

In order to test children's budding theory of mind, Piaget and Inhelder (1956) developed what came to be known as the three mountain problem. Piaget brought children into his laboratory and let them explore a large three-dimensional model of three mountains, decorated with houses, snow, and other objects. Then, a doll would be placed at a particular location facing the model, and Piaget would present the child with photographs of different vantage points. Children were asked to pick the photograph that matched what the doll could see. A child who was still egocentric would select the photograph that matched what *they* could see, while a child with theory of mind would move around the model to the doll's location in order to answer the question.

As research into autism spectrum disorder (ASD) began to increase, a new test for perspective taking was developed. Two of the hallmark features of individuals on the autism spectrum are difficulty with reading social cues and perspective taking. ASD researchers Baron-Cohen, Leslie, and Frith (1985) examined the interaction between ASD and theory of mind, given the predicted relationship between the two. In their famous study that has been cited over 8,500 times, Baron-Cohen et al. lay out what they call the Sally–Anne test. In it, children with autism, children with Down syndrome, and neurotypically developing children watched a puppet show about two girls, Sally and Anne. Sally has a marble, which the children watch her put into a basket and then leave. Anne removes the marble from the basket and puts it into a nearby box. Sally returns to the scene shortly after, and the children are asked, "Where Sally will look for her marble?"

The results may surprise you. While all three groups of children could tell the experimenter where the marble *actually* was, and even where the marble *originally* was, only 20% of the children with ASD were able to put themselves into Sally's shoes, compared to 84% and 85% of the neurotypical and Down syndrome children, respectively. The authors made a clear case that the results were not due to differences in intelligence—as the ASD group's IQ score was average and yet the group with known intellectual disability, the Down syndrome group, still performed better. It seemed that there was some inherent deficit at a social level—theory of mind. The children in the ASD group had a fundamental difficulty separating themselves from Sally in order to report on what Sally knew.

In order to address this biological difference in perspective taking, one form of therapy for children on the spectrum involves teaching them to associate facial expressions, tone of voice, and other nonverbal social cues with particular internal states. Results using this type of therapy have been mixed, although through experience and continued practice, many children on the spectrum are able to improve their social cognition and, as teens and adults, engage in very socially proficient ways with others (Begeer et al., 2011; Chin & Bernard-Opitz, 2000).

YOUR TURN!

Of all the tests designed to measure animal consciousness, the mirror mark test for self-awareness is the easiest to replicate at home. Rather than observing the effects of marking, however, this research idea focuses on the skill of developing and testing control conditions, first introduced in detail in the "Animal Spotlight" on Happy. A control condition is a situation the animal experiences that is identical in all ways to the experimental condition except for the manipulation itself. Control conditions whittle down what exactly it is that drives a behavior.

To complete this ministudy, you'll need to be familiar with ethograms, which we covered in Chapter 2, Theoretical and Methodological Approaches to Animal Cognition. You'll also need a few flat surfaces of similar size, one of which needs to be a mirror. The mirror will be the experimental condition, and the other surfaces will be control conditions. What you are doing here is first noting if your mirror—the experimental condition—changes behavior, and then, if it does, you are trying to make sure the mirror is *exactly* what is causing the change in behavior. For example, you might think your bird is really excited to see you come home from work because she flaps and squawks—until you notice that there happens to be a cat walking by on the sidewalk just as you open the front door. So, is your bird responding to you or to a potential threat? If presenting the mirror changes your animal's behavior, what alternative explanations do you have to rule out that might change the behavior? Maybe it's just seeing something shiny at all. Aluminum foil is shiny like a mirror, but it does not cause a reflection. Maybe seeing the eyes of another animal is what is causing the change. How could you present the eyes of another animal but not a reflection? Try a front-facing color photo of the species (or breed) being tested, and attach it to the aluminum foil. Maybe the simple introduction of an object that the animal is being asked to interact with is what's driving changes in behavior. In this case, a similar-sized, dull surface-like poster board could do the trick. These are just a few suggestions. You'll want to have one mirror (i.e., experimental) condition and two control conditions you're interested in testing out.

Start with the least engaging of your control conditions, and set it up in a space where your animal subject can interact with it. Set a timer for 1 minute, and build an ethogram with frequency counts for the behaviors you observe. This might include "sniffing," "erect fur," "searching behind," "urinating," "hopping," or "other" (if your animal elects to interact with something else instead of your intervention). We cannot know for sure what you'll see, and behaviors vary from species to species, but hopefully this list gives you an idea. Keep track of the behaviors you observe and how many times they engage in that behavior.

After 1 minute is over, remove the first control condition surface, and move on to the second one. Depending upon the species you are working with, this may mean a day or so goes by or entirely different animals are tested. For example, if you're observing wild deer, avoid directly interacting with them by switching your conditions while they're gone. Whenever you're ready, repeat your 1-minute observation using your original ethogram. Record frequency counts for original behaviors as well as frequency counts for any new behaviors you observe in this second condition. Finally, move on to your mirror condition, and do a third round of observation. If you see signs of stress or aggression in your animal, we encourage you to stop if you can. One second of aggression is all you need to be able to conclude "behavior changed" without unnecessarily stressing out your subject for a full minute.

Organize your data, and look for themes and patterns. Did certain behaviors on your ethogram increase or decrease, did entirely new behaviors emerge, or is there a story you can tell using your data that provides answers about what your animal might be experiencing internally as it moves through the conditions? In reviewing your data, also consider the main criticism of the mirror test. If your animal is not a member of a visually guided species, you may have recorded a lot of "other" behaviors. An animal that does not seem to care about even the mirror may be sensitive more to smells and sounds than visuals. In addition to a lesson on developing and implementing control conditions, this ministudy may also offer real-life experience with how important it is to match the task to the species.

REFERENCES

Accreditation Standards and Related Policies. (2017). Retrieved from https://www.aza.org/assets/2332/aza-accreditation-standards.pdf

Almli, L. M., & Burghardt, G. M. (2006). Environmental enrichment alters the behavioral profile of ratsnakes (*Elaphe*). *Journal of Applied Animal Welfare Sciences, 9*(2), 85–109. doi:10.1207/s15327604jaws0902

Amsterdam, B. (1972). Mirror self-image reactions before age two. *Developmental Psychobiology, 5*(4), 297–305. doi:10.1002/dev.420050403

Anderson, J. R., & Gallup, G. G. (2015). Mirror self-recognition: A review and critique of attempts to promote and engineer self-recognition in primates. *Primates, 56*, 317–326. doi:10.1007/s10329-015-0488-9

Andics, A., Gábor, A., Gácsi, M., Faragó, T., Szabó, D., & Miklósi, Á. (2016). Neural mechanisms for lexical processing in dogs. *Science, 353*, 1030–1032. doi:10.1126/science.aaf3777

Ari, C., & D'Agostino, D. P. (2016). Contingency checking and self-directed behaviors in giant manta rays: Do elasmobranchs have self-awareness? *Journal of Ethology, 34*, 167–174. doi:10.1007/s10164-016-0462-z

Bacqué-Cazenave, J., Cattaert, D., Delbecque, J. P., & Fossat, P. (2017). Social harassment induces anxiety-like behaviour in crayfish. *Scientific Reports, 7*, 39935. doi:10.1038/srep39935

Baracchi, D., Lihoreau, M., & Giurga, M. (2017). Do insects have emotions? Some insights from bumble bees. *Frontiers in Behavioral Neuroscience, 11*, 157. doi:10.3389/fnbeh.2017.00157

Baragli, P., Demuru, E., Scopa, C., & Palagi, E. (2017). Are horses capable of mirror self-recognition? A pilot study. *PLoS One, 12,* e0176717. doi:10.1371/journal.pone.0176717

Baron-Cohen, S., Leslie, A. M., & Frith, U. (1985). Does the autistic child have a "theory of mind"? *Cognition, 21,* 37–46. doi:10.1016/0010-0277(85)90022-8

Bateson, M., Desire, S., Gartside, S. E., & Wright, G. A. (2011). Agitated honeybees exhibit pessimistic cognitive biases. *Current Biology, 21*(12), 1070–1073. doi:10.1016/j.cub.2011.05.017

Bateson, M., & Matheson, S. M. (2007). Performance on a categorisation task suggests that removal of environmental enrichment induces 'pessimism' in captive European starlings (*Sturnus vulgaris*). *Animal Welfare, 16,* S33–S36. Retrieved from https://www.ufaw.org.uk/the-ufaw-journal/animal-welfare

Bateson, M., & Nettle, D. (2015). Development of a cognitive bias methodology for measuring low mood in chimpanzees. *PeerJ, 3,* e998. doi:10.7717/peerj.998

Batsching, S., Wolf, R., & Heisenberg, M. (2016). Inescapable stress changes walking behavior in flies—Learned helplessness revisited. *PLoS One, 11,* e0167066. doi:10.1371/journal.pone.0167066

Begeer, S., Gevers, C., Clifford, P., Verhoeve, M., Kat, K., Hoddenbach, E., & Boer, F. (2011). Theory of mind training in children with autism: A randomized controlled trial. *Journal of Autism and Developmental Disorders, 41*(8), 997–1006. doi:10.1007/s10803-010-1121-9

Beran, M. J., Perdue, B. M., Church, B. A., & Smith, J. D. (2016). Capuchin monkeys (*Cebus paella*) modulate their use of an uncertainty response depending on risk. *Journal of Experimental Psychology: Animal Learning and Cognition, 42,* 32–43. doi:10.1037/xan0000080

Beran, M. J., & Smith, J. D. (2012). Information seeking by rhesus monkeys (*Macaca mulatta*) and capuchin monkeys (*Cebus apella*). *Cognition, 120,* 90–105. doi:10.1016/j.cognition.2011.02.016

Bliss-Moreau, E. (2017). Constructing nonhuman animal emotion. *Current Opinion in Psychology, 17,* 184–188. doi:10.1016/j.copsyc.2017.07.011

Brandl, J. L. (2018). The puzzle of mirror self-recognition. *Phenomenology and the Cognitive Sciences, 17,* 279–304. doi:10.1007/s11097-016-9486-7

Bräuer, J. (2014). What dogs understand about humans. In J. Kaminski & S. Marshall-Pescini, *The social dog: Behaviour and cognition* (pp. 295–317). San Diego, CA: Academic Press.

Bugnyar, T., Reber, S. A., & Buckner, C. (2016). Ravens attribute visual access to unseen competitors. *Nature Communications, 7,* 10506. doi:10.1038/ncomms10506

Byrne, R. W., Bates, L. A., & Moss, C. J. (2009). Elephant cognition in primate perspective. *Comparative Cognition & Behavior Reviews, 4,* 65–79. doi:10.3819/ccbr.2009.40009

Cabanac, M. (1999). Emotion and phylogeny. *Japanese Journal of Physiology, 49,* 1–10. doi:10.2170/jjphysiol.49.1

Call, J. (2010). Do apes know that they could be wrong? *Animal Cognition, 13*, 689–700. doi:10.1007/s10071-010-0317-x

Call, J., Hare, B., Carpenter, M., & Tomasello, M. (2004). "Unwilling" versus "unable": Chimpanzees' understanding of human intentional action. *Developmental Science, 7*, 488–498. doi:10.1111/j.1467-7687.2004.00368.x

Call, J., & Tomasello, M. (2008). Does the chimpanzee have a theory of mind? 30 years later. *Trends in Cognitive Sciences, 12*, 187–192. doi:10.1016/j.tics.2008.02.010

Cammaerts, M.-C., & Cammaerts, R. (2015). Are ants (*Hymenoptera, Formicidae*) capable of self-recognition? *Journal of Science, 5*, 521–532.

Case, B. C., Lewbart, G. A., & Doerr, P. D. (2005). The physiological and behavioural impacts of and preference for an enriched environment in the eastern box turtle (*Terrapene carolina carolina*). *Applied Animal Behaviour Science, 92*(4), 353–365. doi:10.1016/j.applanim.2004.11.011

Chalmers, D. J. (1996). *The conscious mind: In search of a fundamental theory.* New York, NY: Oxford University Press.

Chapman, H. A., Kim, D. A., Susskind, J. M., & Anderson, A. K. (2009). In bad taste: Evidence for the oral origins of moral disgust. *Science, 323*, 1222–1226. doi:10.1126/science.1165565

Chin, H. Y., & Bernard-Opitz, V. (2000). Teaching conversational skills to children with autism: Effect on the development of a theory of mind. *Journal of Autism and Developmental Disorders, 30*(6), 569–583. doi:10.1023/A:1005639427185

Clary, D., & Kelly, D. M. (2016). Graded mirror self-recognition by Clark's nutcrackers. *Scientific Reports, 6*, 36459. doi:10.1038/srep36459

Clegg, I. L. K., Rödel, H. G., Boivin, X., & Delfour, F. (2018). Looking forward to interacting with their caretakers: Dolphins' anticipatory behavior indicates motivation to participate in specific events. *Applied Animal Behaviour Science, 202*, 85–93. doi:10.1016/j.applanim.2018.01.015

Clegg, I. L. K., Rödel, H. G., & Delfour, F. (2017). Bottlenose dolphins engaging in more social affiliative behavior judge ambiguous cues more optimistically. *Behavioural Brain Research, 322*, 115–122. doi:10.1016/j.bbr.2017.01.026

Dally, J. M., Emery, N. J., & Clayton, N. S. (2006). Food-caching Western scrub-jays keep track of who was watching when. *Science, 312*, 1662–1665. doi:10.1126/science.1126539

Dally, J. M., Emery, N. J., & Clayton, N. S. (2010). Avian Theory of Mind and counter espionage by food-caching western scrub-jays (*Aphelocoma californica*). *European Journal of Developmental Psychology, 7*, 17–37. doi:10.1080/17405620802571711

Damasio, A. R. (1995). *L'erreur de Descartes.* Paris: Éditions Odile Jacob.

Darwin, C. R. (1872). *The expression of the emotions in man and animals* (1st ed.). London, UK: John Murray.

Deakin, A., Mendl, M., Browne, W. J., Paul, E. S., & Hodge, J. J. L. (2018). State-dependent judgement bias in *Drosophila*: Evidence for evolutionarily primitive affective processes. *Biology Letters, 14*, 20170779. doi:10.1098/rsbl.2017.0779

Delfour, F., & Marten, K. (2001). Mirror image processing in three marine mammal species: Killer whales (*Orchinus orca*), false killer whales (*Pseudorca crassidens*) and California sea lions (*Zalophus californianus*). *Behavioural Processes, 53*, 181–190. doi:10.1016/S0376-6357(01)00134-6

Devaine, M., San-Galli, A., Trapanese, C., Bardino, G., Hano, C., Jalme, M. S., … Daunizeau, J. (2017). Reading wild minds: A computational assay of theory of mind sophistication across seven primate species. *PLOS Computational Biology, 13*, e1005833. doi:10.1371/journal.pcbi.1005917

de Waal, F. M. B., Dindo, M., Freeman, C. A., & Hall, M. J. (2005). The monkey in the mirror: Hardly a stranger. *Proceedings of the National Academy of Sciences, 102*, 11140–11147. doi:10.1073/pnas.0503935102

Doyle, R. E., Fisher, A. D., Hinch, G. N., Boissy, A., & Lee, C. (2010). Release from restraint generates a positive judgement bias in sheep. *Applied Animal Behaviour Science, 122*, 28–34. doi:10.1016/j.applanim.2009.11.003

Gallup, G. G. (1970). Chimpanzees: Self-recognition. *Science, 167*, 86–87. doi:10.1126/science.167.3914.86

Gallup, G. G. (1982). Self-awareness and the emergence of mind in primates. *American Journal of Primatology, 2*, 237–248. doi:10.1002/ajp.1350020302

Gallup, G. G., & Platek, S. M. (2002). Cognitive empathy presupposes self-awareness: Evidence from phylogeny, ontogeny, neuropsychology, and mental illness. *Behavioral and Brain Sciences, 25*(1), 36–37. doi:10.1017/S0140525X02380014

Feinstein, J. S., Adolphs, R., Damasio, A., & Tranel, D. (2011). The human amygdala and the induction and experience of fear. *Current Biology, 21*, 34–38. doi:10.1016/j.cub.2010.11.042

Fraser, D. (2009). Assessing animal welfare: Different philosophies, different scientific approaches. *Zoo Biology, 28*(6), 507–518. . doi:10.1002/zoo.20253

Harding, E. J., Paul, E. S., & Mendl, M. (2004). Cognitive bias and affective state. *Nature, 427*, 312. doi:10.1038/427312a

Henry, S., Fureix, C., Rowberry, R., Bateson, M., & Hausberger, M. (2017). Do horses with poor welfare show 'pessimistic' cognitive biases? *The Science of Nature, 104*, 8. doi:10.1007/s00114-016-1429-1

Hess, U., & Thibault, P. (2009). Darwin and emotional expression. *American Psychology, 64*, 120–128. doi:10.1037/a0013386

Hill, S. P., & Broom, D. M. (2009). Measuring zoo animal welfare: Theory and practice. *Zoo Biology, 28*(6), 531–544. doi:10.1002/zoo.20276

Horowitz, A. (2011). Theory of mind in dogs? Examining method and concept. *Learning & Behavior, 39*, 314–317. doi:10.3758/s13420-011-0041-7

Horowitz, A. (2017). Smelling themselves: Dogs investigate their own odours longer when modified in an "olfactory mirror" test. *Behavioural Processes, 143*, 17–24. doi:10.1016/j.beproc.2017.08.001

Hotta, T., Komiyama, S., & Kohda, M. (2018). A social cichlid fish failed to pass the mark test. *Animal Cognition, 21*, 127–136. doi:10.1007/s10071-017-1146-y

Insabato, A., Pannunzi, M., & Deco, G. (2016). Neural correlates of metacognition: A critical perspective on current tasks. *Neuroscience & Biobehavioral Reviews, 71*, 167–175. doi:10.1016/j.neubiorev.2016.08.030

Iwasaki, S., Watanabe, S., & Fujita, K. (2018). Pigeons (*Columba livia*) know when they will need hints: Prospective metacognition for reference memory? *Animal Cognition, 21*, 207–217. doi:10.1007/s10071-017-1153-z

Kim, H., Hudetz, A. G., Lee, J., Mashour, G. A., Lee, U., & ReCCognition Study Group (2018). Estimating the integrated information measure phi from high-density electroencephalography during states of consciousness in humans. *Frontiers in Human Neuroscience, 12*, 42. doi:10.3389/fnhum.2018.00042

Kirk, C. R., McMillan, N., & Roberts, W. A. (2014). Rats respond for information: Metacognition in a rodent? *Journal of Experimental Psychology: Animal Learning and Cognition, 40*, 249–259. doi:10.1037/xan0000018

Kirkpatrick, C. (2007). Tactical deception and the great apes: Insight into the question of theory of mind. *The University of Western Ontario Journal of Anthropology, 15*, 31–37.

Krupenye, C., Kano, F., Hirata, S., Call, J., & Tomasello, M. (2016). Great apes anticipate that other individuals will act according to false beliefs. *Science, 354*, 110–114. doi:10.1126/science.aaf8110

Kurtycz, L. M., Wagner, K. E., & Ross, S. R. (2014). The choice to access outdoor areas affects the behavior of great apes. *Journal of Applied Animal Welfare Science, 17*(3), 185–197. doi:10.1080/10888705.2014.896213

LeDoux, J. E. (2012). Rethinking the emotional brain. *Neuron, 73*, 653–679. doi:10.1016/j.neuron.2012.02.004

Leung, E. H., & Rheingold, H. L. (1981). Development of pointing as a social gesture. *Developmental Psychology, 17*(2), 215–220. doi:10.1037/0012-1649.17.2.215

Ma, X., Jin, Y., Luo, B., Zhang, G., Wei, R., & Liu, D. (2015). Giant pandas failed to show mirror self-recognition. *Animal Cognition, 18*, 713–721. doi:10.1007/s10071-015-0838-4

Maestripieri, D., Schino, G., Aureli, F., & Troisi, P. (1992). A modest proposal: Displacement activities as an indicator of emotions in primates. *Animal Behaviour, 44*, 967–979. doi:10.1016/S0003-3472(05)80592-5

Maginnity, M. E., & Grace, R. C. (2014). Visual perspective taking by dogs (*Canis familiaris*) in a Guesser-Knower task: Evidence for a canine theory of mind? *Animal Cognition, 17*, 1375–1392. doi:10.1007/s10071-014-0773-9

Markowitz, H. (1982). *Behavioral Enrichment in the Zoo*. New York, NY: Van Nostrand Reinhold.

Markowitz, H., & Stevens, V. J. (1978). *Behavior of captive wild animals*. Chicago, IL: Nelson-Hall.

McMahon, S., Macpherson, K., & Roberts, W. A. (2010). Dogs choose a human informant: Metacognition in canines. *Behavioural Processes, 85*, 293–298. doi:10.1016/j.beproc.2010.07.014

McPhee, M. E. (2002). Intact carcass as enrichment for large felids: Effects on on- and off- exhibit behaviors. *Zoo Biology, 21*, 37–47. doi:10.1002/zoo.10033

Medina, F. S. S., Taylor, A. H. H., Hunt, G. R., & Gray, R. D. (2011). New caledonian crows' responses to mirrors. *Animal Behaviour, 82*(5), 981–993. doi:10.1016/j.anbehav.2011.07.033

Melfi, V. A. (2009). There are big gaps in our knowledge, and thus approach, to zoo animal welfare: A case for evidence-based zoo animal management. *Zoo Biology, 28*(6), 574–588. doi:10.1002/zoo.20288

Mellen, J. D., Markowitz, H., & Stevents, V. J. (1981). Environmental enrichment for Servals, Indian elephants and Canadian otters at Washington Park Zoo, Portland. *International Zoo Yearbook, 21*(1), 196–201. doi:10.1111/j.1748-1090.1981.tb01981.x

Mendl, M., Brooks, J., Basse, C., Burman, O., Paul, E., Blackwell, E., & Casey, R. (2010). Dogs showing separation-related behaviour exhibit a 'pessimistic' cognitive bias. *Current Biology, 20*, R839–R840. doi:10.1016/j.cub.2010.08.030

Metcalfe, J., & Schwartz, B. L. (2016). The ghost in the machine: Self-reflective consciousness and the neuroscience of metacognition. In J. Dunlosky & S. K. Tauber (Eds.), *Oxford handbook of metamemory* (pp. 407–437). Oxford, UK: Oxford University Press.

Miyamoto, K., Osada, T., Setsuie, R., Takeda, M., Tamura, K., Adachi, Y., & Miyashita, Y. (2017). Causal neural network of metamemory for retrospection in primates. *Science, 355*, 188–193. doi:10.1126/science.aal0162

Morgan, G., Kornell, N., Kornblum, T., & Terrace, H. S. (2014). Retrospective and prospective metacognitive judgments in rhesus macaques (*Macaca mulatta*). *Animal Cognition, 17*, 249–257. doi:10.1007/s10071-013-0657-4

Nagel, T. (1974). What is it like to be a bat? *The Philosophical Review, 83*, 43–450. doi:10.2307/2183914

Nelder, K., Collier-Baker, E., & Nielson, M. (2015). Chimpanzees (*Pan troglodytes*) and human children (*Homo sapiens*) know when they are ignorant about the location of food. *Animal Cognition, 18*, 683–699. doi:10.1007/s10071-015-0836-6

Osvath, M., & Karvonen, E. (2012). Spontaneous innovation for future deception in a male chimpanzee. *PLoS One, 7*, e36782. doi:10.1371/journal.pone.0036782

Panksepp, J. (2005). Affective consciousness: Core emotional feelings in animals and humans. *Consciousness and Cognition, 14*, 30–80. doi:10.1016/j.concog.2004.10.004

Panksepp, J., & Burgdorf, J. (2003). "Laughing" rats and the evolutionary antecedents of human joy? *Physiology and Behavior, 79*, 533–547.

Patzke, N., Olaleye, O., Haagensen, M., Hof, P. R., Ihunwo, A. O., & Manger, P. R. (2014). Organization and chemical neuroanatomy of the elephant (*Loxodonta Africana*) hippocampus. *Brain Structure and Function, 219,* 1587–1601. doi:10.1007/s00429-013-0587-6

Paukner, A., Anderson, J. R., & Fujita, K. (2006). Redundant food searches by capuchin monkeys (*Cebus apella*): A failure of metacognition? *Animal Cognition, 9,* 110–117. doi:10.1007/s10071-005-0007-2

Paul, E. S., Harding, E. J., & Mendl, M. (2005). Measuring emotional processes in animals: The utility of a cognitive approach. *Neuroscience and Biobehavioral Reviews, 29,* 469–491. doi:10.1016/j.neubiorev.2005.01.002

Pepperberg, I. M., Garcia, S. E., Jackson, E. C., & Marconi, S. (1995). Mirror use by African Grey parrots (*Psittacus erithacus*). *Journal of Comparative Psychology, 109*(2), 182–195. doi:10.1037/0735-7036.109.2.182

Perry, C. J., & Baciadonna, L. (2017). Studying emotion in invertebrates: What has been done, what can be measured and what they can provide. *Journal of Experimental Biology, 220,* 3856–3868. doi:10.1242/jeb.151308

Perry, C. J., Baciadonna, L., & Chittka, L. (2016). Unexpected rewards induce dopamine-dependent positive emotion-like state changes in bumblebees. *Science, 353,* 1529–1531. doi:10.1126/science.aaf4454

Perry, C. J., & Barron, A. B. (2013). Honey bees selectively avoid difficult choices. *Proceedings of the National Academy of Sciences, 110,* 19155–19159. doi:10.1073/pnas.1314571110

Piaget, J. (1953). *Origin of intelligence in the child.* London, UK: Routledge and Kegan Paul.

Piaget, J., & Inhelder, B. (1956). *The child's conception of space.* London, UK: Routledge.

Plotnik, J. M., de Waal, F. B. M., & Reiss, D. (2006). Self-recognition in an Asian elephant. *Proceedings of the National Academy of Sciences, 103,* 17053–17057. doi:10.1073/pnas.0608062103

Plotnik, J. M., Lair, R., Suphachoksahakun, W., & de Waal, F. B. M. (2011). Elephants know when they need a helping trunk in a cooperative task. *Proceedings of the National Academy of Sciences of the United States of America, 108,* 5116–5121. doi:10.1073/pnas.1101765108

Povinelli, D. J., & Vonk, J. (2004). We don't need a microscope to explore the chimpanzee's mind. *Mind & Language, 19,* 1–28. doi:10.1111/j.1468-0017.2004.00244.x

Premack, D., & Woodruff, G. (1978). Does the chimpanzee have a theory of mind? *Behavioral and Brain Sciences, 4,* 515–526. doi:10.1017/S0140525X00076512

Prior, H., Schwarz, A., & Güntürkün, O. (2008). Mirror-induced behavior in the magpie (*Pica pica*): Evidence for self-recognition. *PLoS Biology, 6,* 1642–1650. doi:10.1371/journal.pbio.0060202

Proctor, H. S., & Carder, G. (2015). Measuring positive emotions in cows: Do visible eye whites tell us anything? *Physiology & Behavior, 147,* 1–6. doi:10.1016/j.physbeh.2015.04.011

Reefmann, N., Wechsler, B., & Gygax, L. (2009). Behavioural and physiological assessment of positive and negative emotion in sheep. *Animal Behaviour, 78*, 651–659. doi:10.1016/j.anbehav.2009.06.015

Reiss, D., & Marino, L. (2001). Mirror self-recognition in the bottlenose dolphin: A case of cognitive convergence. *Proceedings of the National Academy of Sciences, 98*, 5937–5942. doi:10.1073/pnas.101086398

Rochat, P. (2003). Five levels of self-awareness as they unfold early in life. *Consciousness and Cognition, 12*, 717–731. doi:10.1016/S1053-8100(03)00081-3

Roma, P. G., Silberberg, A., Huntsberry, M. E., Christensen, C. J., Ruggiero, A. M., & Suomi, S. J. (2007). Mark Tests for mirror self-recognition in capuchin monkeys (*Cebus apella*) trained to touch marks. *American Journal of Primatology, 69*, 989–1000. doi:10.1002/ajp.20404

Rosati, A. G., & Santos, L. R. (2016). Spontaneous metacognition in rhesus monkeys. *Psychological Science, 27*, 1181–1191. doi:10.1177/0956797616653737

Seligman, M. E., & Maier, S. F. (1967). Failure to escape traumatic shock. *Journal of Experimental Psychology, 74*, 1–9. doi:10.1037/h0024514

Shepherdson, D. J., Mellen, J. D., & Hutchins, M. (1988). *Second nature: Environmental enrichment for captive animals*. Washington, DC: Smithsonian Institution Press.

Smith, J. D., Shields, W. E., Schull, J., & Washburn, D. A. (1997). The uncertain response in humans and animals. *Cognition, 62*, 75–77. doi:10.1016/S0010-0277(96)00726-3

Soler, M., Pérez-Contreras, T., & Peralta-Sánchez, J. M. (2014). Mirror-mark tests performed on jackdaws reveal potential methodological problems in the use of stickers in avian mark-test studies. *PLoS One, 9*(1), e86193. doi:10.1371/journal.pone.0086193

Tan, H. M., Ong, S. M., Langat, G., Bahaman, A. R., Sharma, R. S. K., & Sumita, S. (2013). The influence of enclosure design on diurnal activity and stereotypic behaviour in captive Malayan Sun bears (*Helarctos malayanus*). *Research in Veterinary Science, 94*(2), 228–239. doi:10.1016/J.RVSC.2012.09.024

Tarou, L. R., & Bashaw, M. J. (2007). Maximizing the effectiveness of environmental enrichment: Suggestions from the experimental analysis of behavior. *Applied Animal Behaviour Science, 102*(3–4), 189–204. doi:10.1016/j.applanim.2006.05.026

Templer, V. L., Lee, K. A., & Preson, A. J. (2017). Rats know when they remember: Transfer of metacognitive responding across odor-based delayed match-to-sample tests. *Animal Cognition, 20*, 891–906. doi:10.1007/s10071-017-1109-3

Tononi, G. (2004). An information integration theory of consciousness. *BMC Neuroscience, 5*, 42. doi:10.1186/1471-2202-5-42

Udell, M., & Wynne, C. (2011). Reevaluating canine perspective-taking behavior. *Learning & Behavior, 39*, 318–323. doi:10.3758/s13420-011-0043-5 van der Vaart, E., Verbrugge, R., & Hemelrijk, C. K. (2012). Corvid re-caching without 'theory of mind': A model. *PLoS One, 7*, e32904. doi:10.1371/journal.pone.0032904

Walker, J. K., McGrath, N., Handel, I. G., Waran, N. K., & Phillips, C. J. C. (2014). Does owning a companion animal influence the belief that animals experience emotions such as grief? *Animal Welfare, 23*, 71–79. doi:10.7120/09627286.23.1.071

Watanabe, A., & Clayton, N. S. (2016). Hint-seeking behavior of Western scrub-jays in a metacognition task. *Animal Cognition, 19*, 53–64. doi:10.1007/s10071-015-0912-y

Communication Between Animals

At a very basic level, communication involves passing an information-containing signal from a sender to a receiver (Seyfarth & Cheney, 2017). Though we tend to think of communication as a sophisticated, highly complex process, a great deal of human and nonhuman communication occurs without a hint of cognitive effort. For example, animals who are brightly colored and broadcast to the world "I'm poisonous; don't eat me!" communicate critically important information without even trying. Much discussion, and the vast majority of this chapter, revolves around these kinds of honest signals—ones that provide true information—being communicated between or within species (Seyfarth & Cheney, 2003). Whether it is visual, olfactory, or vocal, most species communicate in direct ways that affect the behavior of others.

The viceroy butterfly (*Limenitis archippus*) capitalizes on looking nearly identical to its poisonous orange-and-black cousin, the monarch butterfly (*Danaus plexippus*), though it is actually harmless. The animal kingdom is full of examples like this where species communicate dishonest signals, and where information is conveyed that somehow deceives potential mates, predators, or rivals. We open this chapter with a few of our favorite examples of how animals use communication to deceive one another.

Was anyone else's favorite Golden Book *Little Yip-Yip and His Bark* (also known as "the-one-about-the-dog-that-wasn't-the-Poky-Little-Puppy")? Poor little Yip-Yip got made fun of by all the other animals on the farm because his bark was so tiny it could not scare away anything. But one day, he realized if he barked inside something hollow, his bark sounded bigger (Jackson & Jackson, 1950). A similar phenomenon can be found in some species of frogs (e.g., green frogs, *Rana clamitans*), which will lower the pitch of their calls so that nearby rivals perceive them to be larger than they really are. Larger frogs are less likely to be challenged over territory and are more likely to find mates, making this dishonest strategy adaptive (Bee, Perrill, & Owen, 2000). Taking the frog's behavior a step further, some orangutans incorporate tools to sound bigger and badder. These animals will use their hands, leaves, or a combination of both to emit a "kiss squeak" alarm call that makes them sound larger to both predators and any other orangutans nearby (Hardus, Lameira, Van Schaik, & Wich, 2009).

Food is often at the center of dishonest signaling. Tufted capuchin monkeys will scare other monkeys away from food they find by sounding alarm calls that are reserved for when a predator is around (Wheeler, 2009). Several different types of birds sound their own species' alarm calls to get access to food (Flower, 2011; Møller, 1988), identify the species of bird that has a desired food item and mimic its alarm call so that it flies away (Flower, Gribble, & Ridley, 2014), or skip alarm calling altogether and mimic a predator's vocalizations (Flower, 2011). Rather than scaring others away from food, some hungry female fireflies (*Photuris* spp.) mimic the flashing patterns of other species' females and lure hopeful males to their waiting jaws (Lloyd, 1984). To keep their young from becoming food, some ground-dwelling birds put on Oscar-worthy performances when predators get too close to their nests. By faking the flapping and struggling of a real injured bird, they deceptively lure predators away from their nestlings, then take to the sky just before they are attacked (Chisholm & Pearse, 1936).

When it comes to gaining access to mates, dishonest signaling gets downright dirty. Male chickens will make food calls when no food is present just to draw females closer to their location and increase their chances of mating (Gyger & Marler, 1988). Taking deception a step further, some young male cuttlefish (*Sepia apama*, similar to a squid) intentionally change the physical shape, texture, and color of their bodies to impersonate female cuttlefish, sneak past larger males that are locked in combat with other males, and mate with the very females the larger males are fighting over (Norman, Finn, & Tregenza, 1999). While the extent to which these many examples of dishonest signaling indicate intention or cognition is difficult to know, the animals' outward behavior seems to suggest something more complex than the viceroy butterfly's color mimicry.

As we return to honest signaling, the focus of this chapter, we'll start by considering our five senses: sight, hearing, smell, taste, and touch. The way an animal looks is almost always indicative of its health, though colors can also communicate information about fertility in some animals. Many species of primate females show a distinct, brightly colored genital swelling when they are receptive to mates. Sounds, smells, and tastes (e.g., urine) can also communicate health and fertility, along with territory boundaries, as anyone who has ever tried to walk a dog in a new neighborhood can attest. Scent can be used to distinguish not just specific identity of individuals but age, sex, reproductive status, and even dominance (Eisenberg & Kleiman, 1972). Touch, such as wrestling, patting, biting, or grooming, can communicate information related to dominance, submission, willingness to play, reconciliation, aggression, and much more. For example, elephants greet one another following separation by touching their trunks on each other's bodies (Payne, 2003) and they communicate reassurance after a stressful situation by placing their trunk in the mouth of a worried conspecific (Plotnik & de Waal, 2014).

No matter how complex it might seem at a physiological or social level, most communication does not tell us much about an animal's cognitive abilities. Peacock (*Pavo cristatus*) tails are meant to show off health. A big, bright tail communicates "I am in good health. I can make these pretty pigmented feathers, and I can drag this huge tail around me without being caught by a predator." Peacock tails and information gathered from urine are extremely important ways animals communicate, but they are not particularly cognitive.

When studying animal communication and the extent to which cognition is involved, it's sometimes a challenge to develop scientific interpretations of observations without imposing human-centric assumptions (Kako, 1999; Shettleworth, 2010). I (AK) talk to my dog all the time. He's obviously "telling" me something when he walks over and barks at me or he paws at my arm, right? As is often the case with animals, it depends. Yes, there are dogs with impressive vocabularies for responding to human words (e.g., Pilley & Reid, 2011), but generally speaking, my dog has learned that pawing at my arm leads to being let out. Is this communication? Yes. Is my dog attempting to convey in words, "Mommy, it is sunny and warm outside, and I would like to stretch my legs. Can you please turn this doorknob so I can go out?" Probably not. What my dog is doing is called conditioned learning. When he first came home and he wanted to go out, he would come near me. Maybe once he accidentally bumped me, so I noticed he was there and thought "maybe he needs to go out." In his mind, the bump and going out became associated; next time he wanted to go out, he bumped me. And gradually that bump morphed into pawing at my arm. This type of conditioning is both a learned response and communication, but it does not require higher cognitive abilities on the part of my dog, although he may certainly have and use them for other purposes.

FEATURES OF ADVANCED COMMUNICATION

There are two major communication features that are indicative of higher cognitive abilities rather than simply physiology, reflexes, or basic conditioning. The first is referential signaling, where an abstract (i.e., meaningless) vocalization or gesture stands in the place of an actual object in the real world. For example, the combination of sounds I make when I vocalize "ahh-pull" is arbitrary and meaningless, except in the context of the English language when we are referencing the very real object in the world called "apple." The word "apple" stands in the place of the actual object apple such that when I say "Bring me an apple," another English speaker would know what to go get. The second higher-order communicative feature is syntax, in which vocalizations or gestures can be combined in different ways to create new meanings. For English, think grammar rules like subject-verb-object here. "The boy hit the girl" and "The girl hit the boy" use the same words that reference real objects in the world, but the word order changes the meaning of the information being communicated.

Referential Signaling

Referential signaling is extremely important when studying complex communication systems. If you hear the words "pink elephant," most likely you conjure a pink elephant in your mind. The words "pink elephant" have an immediate reference to the real objects in the world that are the color pink and an elephant. In reality, it's actually very difficult to determine that an animal or even a human is imagining the real-world objects based on their outward behavior, which is mainly what researchers are able to measure. The phrase "pink elephant" would likely make someone giggle even without conjuring up the actual image of a pink elephant in their mind. So, if they giggle upon hearing "pink elephant," it could be because they saw one in their mind or because of the absurdity of the request to imagine one. It would be really hard for you to figure that out without asking that person directly to self-report on why they giggled, and of course, self-report is not possible with animals.

One way we have been able to examine reference in vocal communication systems is to study species that have multiple categories of behavioral triggers associated with a single concept, such as the concept of "predator." Diana monkeys (*Cercopithecus diana*) can be alerted to a predator by the actual vocalization of the predator (e.g., an eagle scream), or an alarm call made by a monkey in the group. If an eagle alarm call is sounded, the group's response is the same—to get out of the treetops. Likewise, upon being alerted to the presence of a leopard, the response is to get off the ground. What we find is that the calls appear to be functionally referential. They are used to refer to something that is not physically present or visible. Upon hearing a monkey's eagle alarm call, the monkeys will leave the treetops. Now, if that alarm call is followed by the cry of the actual eagle, the monkeys show very little reaction—they have already left the treetops and are safe. That lack of reaction suggests the monkeys already have conjured the eagle in their minds. The real eagle's scream does not add any new information and therefore does not warrant any additional reaction. Similarly, these calls can be desensitized. If an eagle scream is played over and over but no eagle ever appears, the monkeys begin to ignore it. They do not bother to leave the treetops anymore. If, after they are desensitized, a monkey's eagle alarm call is played, it will also be ignored. Wolf has already been cried in regard to the eagle, so the eagle alarm call is now just as meaningless as the real eagle cry (Zuberbühler, Cheney, & Seyfarth, 1999).

Chimpanzees, too, seem to share information with others about food and danger and, moreover, seem to be selective with regards to whom they chose to communicate. Referential understanding allows chimpanzees to provide information selectively to others based on knowing who those others are (i.e., their social rank), knowing what they might both need, and to "know what others know" (allowing purposeful deceit). For example, if a subordinate chimpanzee is aware of hidden food and also knows that a dominant chimpanzee is not, it will refrain from communicating about it or retrieving it until the dominant chimpanzee has left the area. For this to be the case, it seems likely that the chimpanzee must be able to hold the concept of food in its mind, even when it is not visible (Schel, Machanda, Townsend, Zuberbühler, & Slocombe, 2013; Schel, Townsend, Machanda, Zuberbühler, & Slocombe, 2013; Schel, Tranquilli, & Zuberbühler, 2009; Seyfarth, Cheney, & Marler, 1980).

Additional evidence for referential signaling comes from some of the stars of the animal cognition world. Throughout the history of the field, several primates have been taught to communicate with humans via American Sign Language (ASL, or modified versions thereof). Two well-known language-trained apes, Washoe and Koko, have been known to combine signs in flexible, complex ways. For example, Washoe combined the signs for WATER and BIRD when asked to label a swan (Fouts, 1975), and Koko combined FINGER and BRACELET for ring (Patterson, 1986; Patterson & Linden, 1981). To do this, it would be necessary to have a concept of both finger and bracelet as objects with their own characteristics, then combine the two in a novel way.

Alex the African grey parrot (*Psittacus erithacus*) has demonstrated similar behavior using speech. He once combined the words "banana" and "cherry" to label an apple as "banerry." Like with Washoe and Koko, Alex would need to have a concept of banana and cherry as separate objects in order to connect them to the similar features found in apples (Pepperberg, Brese, & Harris, 1991). In addition to novel recombinations of words, Alex could also label objects by color, shape, material, or number of items, and he could generalize as he learned the labels. For example, if he learned "red" from a red square, he would subsequently be able to transfer that label to call a red circle "red" as well. He was able to do this with a variety of characteristics and, surprisingly to some, with sets of items (i.e., find all the green blocks, where there were also blue blocks present), showing that he could communicate about arbitrary concepts and categories for objects in the world, similar to the Diana monkeys' concepts of eagles and leopards as predators (Pepperberg, 2013).

SYNTAX

Once referential abilities are established, some species are able to use syntax to change the meaning of a message by manipulating the order of the vocalizations or gestures in the signal they are communicating. Many people are familiar with the sight of the lone sentry prairie dog (*Cynomys gunnisoni*) keeping watch over his colony. What you might find surprising, however, is that his "ahhhhhh!" alarm call is not just a reflexive scream in response to danger. Instead, there is an added complexity to the prairie dog alarm call system that requires a deeper understanding of the type of danger that is approaching. This requires the ability to combine small vocalization units in a specific order to convey a particular meaning, much like we would put words together to form a sentence. Because they appear to understand that their vocalizations refer to particular characteristics, predators, or situations, sentries use the intentional ordering of their vocalizations to share important details with the rest of colony, such as the type, size, speed, and even the direction from which a predator is approaching (Slobodchikoff, 2002; Slobodchikoff, Kiriazis, Fischer, & Creef, 1991).

Many other species use syntax to create new signals with different meanings. For example, Campbell's monkeys (*Cercopithecus campbelli*) have a set of vocalizations that they combine to produce messages about identity, environmental disturbances, and circumstances. In this species, both simple and complex multivocalization calls are used. The monkeys appear to have a base system of "stem" vocalizations to which additional vocalizations are affixed, depending on the situation. For example, a "stem" for danger might have either the call for "eagle" or "falling tree" attached to it (Ouattara, Lemasson, & Zuberbühler, 2009). With even further specificity, Campbell's monkeys in human care do not seem to produce many of the danger calls used by wild ones; however, they do have a call for "human," which their wild counterparts lack (Ouattara, Zuberbühler, N'goran, Gombert, & Lemasson, 2009). Lastly, it also seems that the length of the vocalization matters, too. Short, simple calls are associated with predation and danger, while longer combinations are used to convey identification or social information, which, when you think about it, makes a lot of sense (Coye, Ouattara, Arlet, Lemasson, & Zuberbühler, 2018).

The composition of a songbird's song represents another great example of syntax. Songs contain variation in both the type and order of the vocalizations, and the order matters. Researchers have identified two vocalizations connected to specific meanings in the Japanese great tit (*Parus minor*) communication system. The "ABC" vocalization causes birds to scan the sky for danger, and the "D" vocalization causes the listener to approach the caller. In a laboratory study, birds heard a recording of another bird vocalize "ABC-D," which caused subjects to first scan the area for danger and then approach the caller. This showed the vocalizations have a combined meaning. However, order matters. When subjects heard "D-ABC," they did nothing. It appeared that reversing the order is incorrect syntax and not recognized as meaningful information (Suzuki, Wheatcroft, & Griesser, 2016). It's not an exact match, but think of it like how the words "I" and "jump" have their own separate meanings, as does the sentence "I jump"; however, "jump I" is nonsensical.

The dolphins Phoenix and Akeakamai, who are profiled in greater detail later in this chapter, showed us that dolphins are also sensitive to syntactic commands and will respond appropriately. For example, Phoenix and Akeakamai could understand that SURFBOARD-FETCH-BALL (take the surfboard to the ball) differed from BALL-SURFBOARD-FETCH (take the ball to the surfboard). Akeakamai would also respond appropriately (or not, as the case may be) when asked to do impossible commands. In one study, when she was given the command SPEAKER-FETCH-BALL, which means bring the (mounted, unmovable) underwater speaker to the ball, she took the ball with her to press on a "no" response paddle (Herman & Forestell, 1985; Herman, Richards, & Wolz, 1984). This ability to spontaneously apply the rule of subject-verb-object would make any English teacher proud and strongly suggests the natural dolphin communication system likely also contains rule-based grammars like the prairie dogs' and Campbell's monkeys' communication systems. By studying dolphins' responses to artificial communication systems in the laboratory, researchers can address challenging questions, such as deciphering the complex wild dolphin communication system.

In all the preceding cases, the sounds being combined are believed to refer specifically to objects. There is one well-known case where it's likely that the acoustic units can be combined in different patterns that are meaningful, but in and of themselves, the individual sounds are meaningless, that is, syntax without referential understanding. Humpback whale (*Megaptera novaeangliae*) song is composed of small units (similar to musical notes), which are in turn arranged into larger combinations and even larger themes. These themes are organized into songs composed of specific, consistent patterns. The songs, sung mostly by males and believed to be displays of fitness during courtship, may not have a referential meaning but can be specific to a particular population based on region or genetics (Garland et al., 2011; Handel et al., 2012; Winn & Winn, 1978). There is also evidence that when the whales migrate, song units from one population can be picked up and integrated into the songs of another (Kaufman, Green, Seitz, & Burgess, 2012). The way humpback whale songs are put together—the syntax built from common acoustic units—has allowed for identification of what researchers call dialects (Winn et al., 1981), similar to how one language can vary by geographic region.

TEACHING LANGUAGE TO ANIMALS

While it might not seem like it, the word "language" is highly contentious, especially among human and nonhuman animal researchers. Though definitions and criteria vary widely, a language is a communication system that contains referential signaling, syntax, displacement, and generativity. Displacement means users can communicate about information outside of time and place as well as about abstract information that does not have a real-world tangible connection. Being able to communicate about something that happened in the past, or in a different location, or concepts like justice or peace illustrates the displacement quality of language. By generativity, linguists mean that from a finite set of units, an infinite number of combinations can be made. I (ECW) always find it fascinating to hear a sentence I distinctly know I've never heard before: "A purple dinosaur wearing pajama bottoms crashed through the ceiling at the bank." The fact that you've never heard that sentence before (I think), and yet I was able to communicate such a specific piece of information, showcases the generative quality of language.

All of this to say, while research shows other species have some of the preceding criteria, no species seems to have all of them in quite the same way as humans do. Thus, language can be defined simply as the system humans use to communicate. While it might be tempting to say that two dogs yipping and barking together are using "dog language" or that two color-changing octopuses are communicating with "octopus language," this would not be correct. In the cases described earlier, also keep in mind that though the words or gestures might make us want to say that parrots, for example, can learn language, it's more appropriate to think of it in terms of parrots learning to use familiar sounds and rules of language in highly sophisticated ways, but not full-blown language itself.

During the second half of the 20th century, language was the most controversial and popular topic in animal cognition. Today, there are very few animal language projects (Beran, Parrish, Perdue, & Washbur, 2014). The first successful instance of teaching an animal rules of our own language (or an artificial language we have created) was with a chimpanzee named Washoe. Washoe was raised by Allen and Beatrix Gardner of the University of Nevada, Reno. She was taught ASL and treated as a member of the Gardner household. The Gardners exposed her to normal everyday activities and the appropriate signs to go along with them. Similar to human children, Washoe learned signs rapidly without rigorous "training." Instead, the Gardners (Gardner & Gardner, 1969) claim that she picked up ASL naturally during social interactions with humans.

A second chimpanzee, Nim Chimpsky, had a different ASL experience. Nim spent more time in formal training sessions. These sessions focused more on conditioning, molding (i.e., moving his hands into the proper sign), and modeling, rather than natural interactions (Terrace, Petitto, Sanders, & Bever, 1979). Nim was much less prolific with signing than Washoe. Some hypothesize that Nim may have been unable to succeed to the degree Washoe did because he did not have the same incorporation into a family unit as Washoe did. As the "Human Application" section at the end of the chapter illustrates, immersion in language via social interaction and exposure to the sociocultural factors that contribute to what it means to be "human" are critical for children to develop language. Some believed (and still believe) that these factors contributed to the communicative differences between Washoe and Nim (Fouts, 1974; Hess, 2008; Terrace et al., 1979).

Over the years, some have studied how animals learn to use symbol-based languages made up by the researchers, rather than

ASL. There have been a variety of successful cases of animals learning either an artificial gestural language or one based on pointing to visual symbols (known as lexigrams). One chimpanzee named Lana at Georgia State University's Language Research Center (LRC) was the first to use lexigrams in an artificial language. The language she, and many others after her at the LRC, learned was called Yerkish (Rumbaugh & Gill, 1977). After Lana, Sherman and Austin were the next group of Yerkish users. These chimpanzees lived together, and it was suspected that they even used the abstract Yerkish symbols to communicate with each other. This came in the form of them requesting objects from each other using the Yerkish lexigrams. Important to note, however, there's no evidence that they ever answered those requests, so it's hard to draw any specific conclusions about what that might mean (Savage-Rumbaugh, Rumbaugh, & Boysen, 1978; Savage-Rumbaugh, Rumbaugh, & Fields, 2009). Overall, all three could create two-word sentences and achieved at least some degree of success in understanding abstract concepts, symbols, and syntax (Barón Birchenall, 2016).

Of all the apes at the LRC, Kanzi the bonobo is arguably the most successful Yerkish user (Shanker, Savage-Rumbaugh, & Taylor, 1999). Like Washoe, language was introduced to Kanzi as it would be to a human child, via contextual interactions with his caretakers (Shanker et al., 1999). Not only does Kanzi use the lexigram keyboard to request items, he can label items as well, meaning he can communicate beyond requests for objects to be handed to him, such as asking caretakers to play his favorite games like chase (Sevcik & Savage-Rumbaugh, 1994). In addition to human caretakers, Kanzi was also raised around other apes (Savage-Rumbaugh & Lewin, 1994), a deviation from many animal language practices. Whether Kanzi's status as a dual citizen of bonobo and human life contributed to his extraordinary communicative abilities is unknown.

Though we must be especially careful not to anthropomorphize, we can learn a great deal about the minds of animals by communicating with them using artificial language systems. For example, language-trained animals can and do make choices in their everyday life, which give us insight into their cognitive abilities. It might not seem immediately important for Kanzi to request an orange soda (his favorite drink). However, look at it this way: Kanzi requesting an orange soda means that he can keep an inventory of the drinks he's had in his memory, that he can communicate a preference, and that he knows an appropriate time to request the drink (e.g., meal time, not bedtime). That's extremely informative about his cognitive abilities. As another example, Alex the African grey parrot showed us that he was capable of understanding the concept of "zero," that he could understand embedded clauses (e.g., "find the green ball" requires finding all the balls—not the trucks or dolls—and then picking out the green one, as opposed to the red or blue ones), and that he could provide the total number of a set of specific objects (Pepperberg, 1992, 2012; Pepperberg & Gordon, 2005).

Alex captures our attention because of his uncanny ability to not only replicate speech sounds but use them in rule-based ways that are very similar to language (Pepperberg, 1999). Parrots show an innate ability to mimic the sounds around them. In the wild, this is thought to help create social bonds within flocks and pairs of parrots. Calls are similar within flocks and can be used to gather the group at dusk, and convergence of the features of the calls is common during pair bonding and mating (Hile, Plummer, & Striedter, 2000; Sewall, Young, & Wright, 2016).

Parrots living with humans adopt people as their "flock," so they learn to use the vocalizations of that flock by spending time with humans. Like human children, juvenile parrots even go through a babbling phase where they play with sounds before they learn actual words (Pepperberg et al., 1991). Word learning happens via natural social interaction at home or in more structured training sessions in laboratories (Colbert-White, Covington, & Fragaszy, 2011; Pepperberg, 2006). Pepperberg had great success teaching speech to Alex and other parrots using the model-rival technique. In it, Alex learned new labels by observing a trainer praise and reward a human assistant for correctly labeling objects (Pepperberg, 1981). Using this method Alex learned not just basic vocabulary but even more complex abilities. In addition to some of the previously mentioned ones, Alex could label objects, colors, and shapes (and combinations thereof); used phrases like "I want" and "wanna go"; and understood relational concepts such as "bigger" and "smallest" (Pepperberg, 2006).

While some parrots like Alex are explicitly trained for research purposes, there are also many anecdotes of language-like behavior by pet birds at home, some of which have been transformed into scientific case studies (Colbert-White et al., 2011; Kaufman, Colbert-White, & Burgess, 2013; Colbert-White, Hall, & Fragaszy, 2016). One pet-parrot-turned-research-star is Cosmo the African grey. Colbert-White et al. (2011) showed that Cosmo's vocabulary extended beyond 200 words and phrases and that she used speech in very language-like ways. For example, she monitored the social context she was in and responded accordingly. When Cosmo's owner spoke with an experimenter and ignored Cosmo verbally, she was more likely to request physical affection. When her owner told her no, she repeated her requests; when her owner ignored her requests, Cosmo would get her attention by saying "I love you," then repeat her request only after her owner took the bait and had replied back "I love you" (Colbert-White et al., 2016). And for whatever the anecdotal evidence is worth, I'm fairly certain that my (AK) African grey, Eliza, is intentionally trying to engage me when I walk in the door and she says "How are you?" in my husband's voice, triggering several sentences of detail on the day's events before I realize I'm talking to the bird. Just like Kanzi with Yerkish or a human child with Korean, parrots living among humans absorb the natural communication system of the human "flock" in which they live, illustrating just how flexible parrots are at vocal communication in the wild and in our homes.

GESTURES AND SOCIAL CUES

Humans are constantly affecting each other's behavior using gestures and social cues. We emphasize important points using our hands, follow others' eye gaze to see what they're looking at, and politely decline to taste someone's meal when they grimaced after their first bite. Gestures and other nonverbal social cues act as helpful signals and communicate others' goals and internal state, whether it's, "Wow, look here with me," "I'm bored," or, "This tastes disgusting."

Researchers who teach communication systems to apes often choose to use nonverbal systems such as ASL or visual symbols. This is usually because of apes' limited vocal abilities (although there is some recent research that contradicts this; Fitch, 2000; Fitch, de Boer, Mathur, & Ghazanfar, 2016; Fitch & Reby, 2001). Rather than vocal communication, apes are actually most skilled at using manual gestures to communicate information. In the wild, chimpanzees use a variety of gestures in specific semantic contexts; different gesture types are used for grooming requests, displays of social rank, food sharing, and travel. Very often—but not always—responses are contextually appropriate (Roberts, Vick, & Buchanan-Smith, 2012), which indicates a level of intentionality by the gesturer. This means apes likely use gestures with a goal to change the behavior of another individual. Further, they're consistently used among animals within a social group, so much so that they have been categorized as a relatively sophisticated gestural "vocabulary" (Hobaiter & Byrne, 2014, 2017).

Apes' ability to follow each other's gestures appears to be innately purposeful and understood—meaning, a chimpanzee does not need to learn that when another chimpanzee looks at him and then gestures toward an apple, the individual is communicating with him about that particular object out in the world. There are several ways to know this. Apes will repeat gestures to human caretakers who do not seem to understand what the message is, basically to say "Hey, bozo, I meant the apple, not the celery" (Cartmill & Byrne, 2007). They will even pause and wait with expectancy to see if a receiver reacts, as in, "I'll be right here waiting for that apple..." (Liebal, Pika, Call, & Tomasello, 2004). Just like you would simplify your language when talking to a small child or choose your words carefully if your friend was known to gossip, chimpanzees also show significant difference in the usage of gestural signals based on whether the individual they are attempting to communicate with is facing them (Hostetter, Cantero, & Hopkins, 2001).

In addition to gesture-based communication systems, some animals use their bodies to communicate information that is picked up by social partners. While large manual gestures are easy to observe in the wild, capturing instances of more subtle social cues like eye-gaze following is difficult due to its nuance and time intensiveness. Imagine having to wait around the clock for days, weeks, or months to capture what could be a 1-second-long social interaction! To make life easier, researchers tend to study how animals respond to humans' nonverbal social cues. From these kinds of studies, we now know that many animals, including chimpanzees (Itakura & Tanaka, 1998), South African fur seals (*Arctocephalus pusillus*, Scheumann & Call, 2004), African grey parrots (Giret, Miklósi, Kreutzer, & Bovet, 2009), and bottlenose dolphins (Herman et al., 1999), will follow humans' pointing and/or gazing. Dolphins will even stiffen their bodies, stop swimming, and "point" their rostrum at objects they want humans to attend to with them. They also periodically pause to look back and monitor the human diver's behavior, then return to pointing. Interestingly, Xitco, Gory, and Kuczaj (2001) noted that the dolphins never did the "monitoring" behavior when a human was not in the water with them, and in a follow-up study in 2004, the authors found that the dolphins were very unlikely to point when the human's back was turned and almost never pointed if the human was swimming the other direction. These two studies indicated to Xitco et al. (2004) that the pointing and monitoring behaviors were intentional attempts at social communication by an animal to a human social partner.

Some hypothesize that via generations of selective breeding, domesticated species have become better than wild species at attending to and learning from a broad range of human social cues like pointing and eye gaze (see Hare, Brown, Williamson, & Tomasello, 2002; horses, Maros, Gácsi, & Miklósi, 2008; pigs, *Sus scrofa domestica*, Nawroth & von Borell, 2015). Along with attending to human pointing and eye gaze, dogs, cats, and goats (*Capra aegagrus hircus*) can also follow human-pointing cues to find hidden food (Miklósi, Pongrácz, Lakatos, Topál, & Csányi, 2005).

In comparison to dogs, wolves (*Canis lupus*) tend not to perform as well in human-pointing studies (Miklósi, Kubinyi, Topál, Viranyi, & Csányi, 2003). Since dogs descended from wolves, why might there be differences? Comparative work with multiple canine and primate species has illustrated the cognitive effects of thousands of years of domestication and coevolution with humans. Despite being more closely related to us than wolves, wolves outperform chimpanzees at finding food a human is pointing to, but domesticated dogs are exponentially better than wolves. This has been dubbed the domestication hypothesis (Hare, 2017; Hare et al., 2002). To understand how domestication changes an animal's behavior and cognition, Russian zoologist Dmitry Belyayev selectively bred a group of wild foxes (*Vulpes vulpes*) for 45 generations for an interest in and lack of aggression toward people (Trut, 1999). Along the way, many unintended traits emerged, including the use of juvenile fox vocalizations into adulthood, friendly behaviors like tail wagging, and a sensitivity to human gestures, despite no specific breeding for these qualities. It appeared the selection for docility and interest in humans also increased the foxes' social cognition in interactions with humans (Hare, 2017; Hare et al., 2005). The power of human interaction was also demonstrated on a smaller scale with feral dogs, dogs growing up in animal shelters, and dogs growing up as pets. Across the three groups, the findings showed increasing levels of understanding and use of gestural information communicated by humans, with pet dogs being the most skilled at interpreting human gestures (Viranyi et al., 2008).

In addition to the process of domestication, time spent with humans also seems important; pet dogs tend to outperform shelter dogs on interpreting and using human social cues (Duranton, Bedossa, & Gaunet, 2016). Taken together, the variety of studies on dog–human communication indicates that thousands of years of coevolution has resulted in dogs being particularly attuned to human social cues as well as able to extract meaning and information from them.

INSECT COMMUNICATION

We've already provided a few examples of insect communication systems, like the femme fatale fireflies that mimic other species' flashing signals (Lloyd, 1984). However, there are some insects that have communication systems that are so complex they belong in a section all their own. Bees and ants are highly social, so it should come as no surprise that their use of visual and chemical signaling might be particularly sophisticated.

The honeybee waggle dance is perhaps the most famed example of communication within an animal species, and it is particularly remarkable considering the presumed simplicity of the species. Karl von Frisch first identified the communicative abilities of the dance after observing that bees quickly congregated around new food sources. He began to hypothesize that hive mates were communicating and set up food stations around a large grassy property. When bees found these food stations, von Frisch was able to track how the bees communicated their locations to other colony members (von Frisch, 1967). Forager bees who have located food sources use the waggle dance to communicate information to recruit bees who must go find the source. The information is communicated using extremely precise angles with the sun as a reference point and has been shown to provide details about distance and direction to allow recruits to locate specific food patches (Gould, Henerey, & MacLeod, 1970). Interestingly, not all bees are equally successful at communicating with the waggle dance, but the overall activity of the hive compensates for these terrible dancers who provide bad directions (Preece & Beekman, 2014).

In a somewhat new development, researchers have been tracking bees' use of olfactory communication about food sources as well. In some cases, this results in an information conflict. While the waggle dance visually advertises the location of food publicly, olfactory cues are regarded as more "private" because they are carried by individuals. When a comparison is made, bees are far more likely to follow private olfactory trails to food than public waggle dances, indicating a definite consideration of the circumstances under which the information is obtained and the modality used to communicate it (Grüter, Balbuena, & Farina, 2008).

Like bees, ants use a variety of methods of communication—visual, olfactory, and tactile—in their interactions (Thienen, Metzler, Choe, & Witte, 2014; Wilson, 1958), although olfactory appears to be the most common. Ants leave the nest and wander until they find a food source. Along the way they leave pheromone trails. Once a food source is found, however, the ant retraces its steps and returns to the colony thereby strengthening the trail. More ants follow the stronger trail, creating positive feedback. The correct trail grows stronger, and the incorrect ones grow weaker.

Simple trails to food sources are fairly noncognitive. However, in addition to laying simple trails, individual ants can leave variations that suggest more complex cognitive behavior. Specifically, the concentration of the pheromones applied to trails affects trail detection and its subsequent use (Thienen et al., 2014). For example, if one food source is smaller than another, an ant may put down a weaker pheromone trail, meaning it will disappear sooner with less use. Ants can tell how populated the trails are, and in turn they judge whether a new food source will be needed soon or if more members of the colony should be directed toward a more bountiful one. Pheromones can even trigger separate memories of the route to a food source, doubling up on the navigational information an ant has (Czaczkes, Grüter, & Ratnieks, 2015). Thus, the complexity of what ants can convey with a combination of pheromones and additional stimuli is both vast and far more advanced than one might imagine.

ANIMAL SPOTLIGHT: PHOENIX AND AKEAKAMAI

There are dozens of species of dolphin, with habitats ranging from saltwater to freshwater, oceans to rivers. Dolphins are cetaceans, a group of aquatic mammals that also includes whales and porpoises. Cetaceans are known to have complex vocal communication systems, but due to their physical size and threat level (many are endangered), only a few have been studied extensively in the wild or while in human care. The bottlenose dolphin lives close enough to coastal areas to make field studies convenient, reproduces well in human care, and seems interested in spending time with humans (Clegg, Rödel, Boivin, & Delfour, 2018). For these reasons, bottlenose dolphins have been at the center of vocal communication and other cognitive research for decades. This is the species we are referring to when we say "dolphin."

Dolphins' vocal communication system is composed of as many as a dozen different sound types, including echolocation clicks, tones, twitters, and "burst pulses," but the vast majority are whistles (Díaz López & Shirai, 2009). These sounds are supplemented by and matched to specific communicative gestures such as tail slaps and jaw pops. If you've ever heard a dolphin vocalize, you may have noticed their mouth was open. Do not be fooled; that's not where the sound comes out. Dolphin vocalizations are emitted through their blowholes! This is because unlike terrestrial mammals, cetaceans do not have vocal cords (Madsen, Jensen, Carder, & Ridgway, 2011). Imagine if humans spoke to one another through their noses; this is essentially how dolphins do it.

The dolphin communication system contains dozens of different sound patterns created by different arrangements of the base sound types, and these patterns appear to have distinct functions related to food, social behavior, sexual play, hunting, courtship, discipline, travel and group coordination, and aggression (e.g., Díaz López & Shirai, 2009; Herzing, 1996; Grdiley, Nastasi, Kriesell, & Elwen, 2015; Papale et al., 2017). It's pretty hard to figure out specifically which behaviors match which vocalizations, considering much of dolphin behavior occurs underwater and that we're really just learning to identify the vocalizations with reliability.

Dolphins are scientifically classified as "toothed whales," as opposed to the humpback whales discussed earlier in the chapter, which are baleen whales. Like the humpbacks, dolphins also appear to have geographically distinct dialects, further demonstrating the flexibility and complexity of their communication system (Morisaka, Shinohara, Nakahara, & Akamatsu, 2005). Another toothed whale, the orca (*Orcinus orca*) has at least one call that appears to be universal across all populations, including those who never come in contact with each other (Rehn, Filatova, Durban, & Foote, 2010). The inclusion of a call like this in the repertoire of all orca populations could be indicative of an innate, or inborn, vocalization, which also interests dolphin researchers.

One class of whistle vocalizations that dolphins produce is considered the dolphin equivalent of names. These signature whistles are individually distinct, learned by each animal in the group, and have been studied extensively given the ease with which they can be detected in recordings (Caldwell & Caldwell, 1965). Signature whistles are used for identification, and members of the same social or familial group will exchange whistles when they are out of visual contact (Sayigh, 1999; Sayigh, Esch, Wells, & Janik, 2007). Signature whistles are also copied back and forth between mothers and calves and within groups of young males ("bachelor groups") to increase bonding and group cohesiveness (King, Harley, & Janik, 2014; Quick & Janik, 2012). This aptitude for vocal learning and vocal flexibility makes them ideal for studies of communication, and more specifically, studies on specific cognitive abilities. Because signature whistles are well studied, we can build upon what we know to further study syntax and semantics in dolphins. For example, we can study how the dolphins are behaving when they whistle or the ordering of the vocalizations with which dolphins "introduce" themselves (Janik & Slater, 1998; Quick & Janik, 2012).

As of now, no one has cracked the code of dolphins' vocal communication. At best, we are at the stage of connecting vocalization types to particular behaviors or contexts, such as rising whistles being associated with social behaviors and multiloop whistles being associated with hunting (López, 2011). This is a long and difficult process, especially when dialects get involved. Part of the challenge comes from knowing who said what when a group of gregarious dolphins swims by. The dolphin social group is incredibly dynamic and fast moving, and recording technology is still trying to catch up.

At the beginning of the animal language craze of the second half of the 20th century, wild dolphins were captured and brought to laboratories to study the extent of their language abilities. Like with apes, early attempts involved trying to give dolphins a life that was as close to a human child's and family unit as possible. One such effort was a government-funded project in the 1960s that involved people and dolphins spending time in an underwater house in the Virgin Islands. The "Dolphnarium" project, as it came to be called, sought to teach the dolphins to pronounce human speech sounds and learn language. Unfortunately, also like early ape language research, the project was eventually defunded due to lack of meaningful findings (Herzing, 2016).

Starting in the late 1970s, a new method of studying communication and language in captive dolphins arose from the ashes, so to speak. Female bottlenose dolphins Phoenix (named for the legendary bird symbolizing rebirth) and Akeakamai (from Hawaiian, meaning "lover of wisdom") resided at the Kewalo Basin Marine Mammal Laboratory in Honolulu, Hawaii, run by a professor of psychology, Dr. Louis Herman, at the University of Hawaii. The dolphins began language training at about 2 years of age, and each was taught an artificial language. Phoenix's language was an auditory system. Sounds were created by a computer and assigned to keys on a computer keyboard. They could then be selected in the order required by the experimenter and played in the tank for Phoenix to hear via an underwater speaker. Akeakamai's language was gestural. Signals consisted of arm movements by a signer standing at the tank wall, visible from the waist up. Elaborate controls, included blindfolding Akeakamai's signer who presented the gestural signals and carefully training anyone who observed and recorded both dolphins' responses, were put in place to avoid unintentionally biasing the data (Herman et al., 1984).

There was intentionally considerable overlap of elements between the two languages, and both contained actions, directionality, and objects (both fixed and moveable). In both cases there was a specific syntax: Phoenix's auditory language had a linear word order, subject-verb-object, while Akeakamai's gestural language was inverse, object-subject-verb (Herman, 2010; Herman et al., 1984).

Both dolphins were extremely successful in their respective languages, showing the ability to understand word order and syntax. Both also seemed to understand that the signs comprising their respective systems referenced real objects and actions. Akeakamai eventually learned an "erase" command, which meant something along the lines of "disregard what was just signed," and Phoenix was able to give responses to combinations of multiple actions, such as FRISBEE-FETCH-UNDER-HOOP (Herman & Forestell, 1985; Herman et al., 1984). The dolphins were also, surprisingly, able to learn their communication systems from television screens. When tested, Phoenix and Akeakamai learned gestures from a television at about the same level of fluency as human trainers who were "experienced" (as opposed to novice or expert; Herman, 2010; Herman, Morrel-Samuels, & Pack, 1990; Savage-Rumbaugh, 1986). Learning a communication system from a television screen was particularly intriguing at the time because it's something that had not been demonstrated in other animals—including chimpanzees—and it's also something human children are quite poor at (Deloache et al., 2010).

The work done with Phoenix focused solely on her artificial language abilities, while Akeakamai also participated in vocal mimicry studies (Richards, Wolz, & Herman, 1984). In these studies, Akeakamai was initially trained to comprehend a wide variety of sounds. As her comprehension improved, she became able to reproduce some of the sounds herself, even when pitch or volume was manipulated, which was something she had not been trained on. Soon, Akeakamai moved from mimicry to actually labeling. When presented with an object, she could produce the correct whistle label with accuracy as high as 91% (Richards et al., 1984). Phoenix, not to be completely outdone in this regard, learned to discriminate tunes and generalize the change in tones across different octaves, a difficult skill for even trained musicians (Herman, 2010).

Akeakamai advanced further with her language than Phoenix, working with longer strings of up to five-sign-long commands (Herman et al., 1984). She was also involved in research studies that tested her response to impossible commands, which we introduced earlier. Herman and his team found that in these situations, Akeakamai would improvise and do what she could. So, if she was given a command like SPEAKER-PHOENIX-HOOP-FETCH, for example, she would bring the hoop to Phoenix and ignore the part about transporting the immobile underwater speaker (Herman, Kuczaj, & Holder, 1993).

In addition to language research, Herman and his team investigated other features of communication. For example, they showed that she understood both single pointing gestures and sequences of two different points (Herman et al., 1999; Pack & Herman, 2007). This indicated that she was doing more than just responding reflexively to isolated commands; she was able to integrate multiple pieces of information at once and respond to commands holistically.

Because of their impressive cognitive abilities, Phoenix, Akeakamai, and other dolphins have often been referred to as aquatic apes (Barrett & Würsig, 2014). While they certainly seem to have some similarities, the most fascinating question is not always what an animal can do but how they came to be able to do it. One of the biggest takeaways from comparing communicative and cognitive abilities of apes and dolphins is an understanding of which skills are the most vital for survival and how those skills manifest in different environments. Dolphins and other whales can teach us a great deal about how an aquatic environment might shape different aspects of similarly complex communication systems.

HUMAN APPLICATION: CHILDREN RAISED WITHOUT LANGUAGE—GENIE

What do humans, dolphins, bats, and hummingbirds all have in common? These animals, in addition to elephants, parrots, songbirds, and some seals, are vocal learners (Boughman, 1998; McCowan & Reiss, 1997; Nottebohm, 1972; Poole, Tyack, Stoeger-Horwath, & Watwood, 2005; Sanvito, Galimberti, & Miller, 2007). Unlike a seagull or even a chimpanzee, whose communication systems are considered to be written into their DNA, vocal learners come equipped with all the hardware to learn their species' vocalization system, but the system itself is not intact. Instead, they must learn their communication systems from members of their own species. If a member of their species is not readily available, they can sometimes adopt another species' communication system. This explains why parrots are so good at learning how to vocalize using human speech; they have the general neural hardware to hear and reproduce almost any consistent sounds in their environment, then add them into their vocal repertoire, whether it's an actual word, a creaky door sound, or the family dog's bark. The case of parrots being adopted into humans' social groups is a best-case scenario for a vocal learner who is not exposed to their own species' vocal communication system. Normally, a vocal learner who does not get extensive exposure to their native tongue is doomed to have a completely disrupted vocal repertoire.

Similarly, for humans, the critical period hypothesis for language states that there is a period of time during which language learning must occur, if language is to be learned to native-like competence. This time frame is called the critical period (Lenneberg, 1967). Researchers are divided on when the critical period ends: Some researchers argue that it ends as early as when a child is about 5 years old (Penfield & Roberts, 1959) or as late as the onset of puberty (Johnson & Newport, 1989; Lenneberg, 1967). We know very little about the specifics of this. Our knowledge stems almost entirely from case studies of children who have lived in situations of severe abuse and deprivation and therefore did not experience (typical) exposure to (typical) language (Doupe & Kuhl, 1999). Fortunately, these cases are rare.

With (thankfully) limited opportunities to study how external circumstances disrupt language learning during the critical period in humans, some researchers have turned to investigating what is believed to be a similar critical-period system in songbird song learning. Songbirds can be divided into two types of song learners: open ended and age limited. Open-ended song learners like canaries are able to learn new songs throughout their lifetimes, and exposure to songs after prolonged isolation appears to be sufficient for song learning (Ball & Hulse, 1998; Doupe & Kuhl, 1999). Nearly every other type of songbird is (like humans are with language) an age-limited learner and therefore has a specific period during which song learning is accomplished. If an age-limited songbird does not get adequate exposure to its species' song to learn it, either it never develops the song or it develops atypically. Though there may be others, humans and songbirds have been the main vocal-learning species where this highly limited critical period has been identified.

In 1970, in California, one of the most severe cases of child abuse in history was uncovered with the discovery of Genie, a 13-year-old girl who could neither speak nor comprehend language. Genie's father had a history of domestic abuse and had supposedly never wanted children in the first place (two older children had died in infancy, one likely due to neglect). From the time Genie was born, she was kept in a dark room at the back of the family's home, strapped either to her crib or to a potty chair in the room. Genie's father terrorized her when she misbehaved, by growling at her and scratching her like an aggressive dog. One way she could misbehave was by making sounds. Her father did not tolerate noise in the house (they did not even own a television or radio), and so in addition to never being spoken to, Genie was punished for trying to speak, so severely, in fact, that she learned to be essentially mute (Curtiss, Fromkin, Krashen, Rigler, & Rigler, 1974; Fromkin, Krashen, Curtiss, Rigler, & Rigler, 1974; Rymer, 1993).

After 13 years of horrific treatment, Genie was discovered purely by chance. Her mother had sustained a head injury years prior, which caused worsening blindness. One day, Genie's mother took her with her to apply for disability benefits. Instead of the disability office, the two accidentally walked into the social services office next door, where Genie's appearance and behavior immediately set off red flags. When she was removed from the house, Genie was physically the size of a child half her age, and her cognitive abilities had also been severely stunted.

In addition to physical and cognitive deficiencies, Genie was past the critical period for language development. There are several hypotheses for what it means for humans to miss the critical period for vocal and gestural (i.e., signed) language development (Morford, Grieve-Smith, MacFarlane, Staley, & Waters, 2008). One hypothesis looks at the critical period relative to the child's development. This idea puts language learning and its critical period on a specific timeline in accordance with the growing brain. It holds that brain development, not age, is key to learning language. For example, if one child reaches a specific developmental milestone and another one has not, the two could be the same age but at different linguistic stages (Newport, 1990). A second hypothesis focuses on the amount of cognitive resources available during the critical period for language learning. Think about a time you were listening to someone speak another language. Could you tell where one word ended and the next began? This is called segmentation of the language stream, and the breaks between words only become apparent once you have been exposed to a language. Humans are sensitive to various cues to help segment language. For example, at birth, humans can discriminate the different phonemes (smallest meaningful units of sound) of *all* the world's languages (Eimas, 1985; Kuhl, 2004). At around 6 months, human infants begin to lose the ability to discriminate the phonemes of all the world's languages and begin to hone in on the phonemes that describe the language(s) they were exposed to (Werker & Tees, 1984). Once infants know which phonemes describe their language(s), they can use this ability, among others, to segment the speech stream and identify what sounds make up words in their language(s) (Thiessen & Saffran, 1993). According to Kuhl et al.'s Native Language Magnet Theory (2008), infants master the difficulties and intricacies of what will become their native language(s) once they begin honing in on the features of their native language(s). Putting these two hypotheses together, the abuse and neglect Genie sustained would have created a perfect storm; her brain's general development was disrupted, and she was never exposed to what could have become for her a native language. As a result, the areas associated with language in her brain did not develop to support language acquisition, and some language-learning abilities were lost for good.

CHAPTER 4 COMMUNICATION BETWEEN ANIMALS

Developmental psychologists and linguists alike were drawn to Genie's case as a rare opportunity to study language acquisition. Under the supervision of professionals, Genie began to communicate using language-like features for the first time, though it became quickly clear that what she was able to learn was far different from normally developing children. Unsurprisingly, her language comprehension was better than her language production. That is, like a very young child, she understood more than she could actually say. The main difficulty seemed to be physiological. Genie appeared to struggle to move her facial muscles in the correct way to form sounds, and doing so seemed to take a good deal of energy and concentration. Imitation was far easier. It was possible for Genie to make the sounds of the human language. She was able to imitate whatever sound was demonstrated to her. However, her spontaneous speech—the words she chose to use on her own—was limited and was generally composed of the sounds that were easy for her to make (Fromkin et al., 1974). After about 5 months in the hospital, Genie began to produce single words spontaneously, meaning without prompting or questioning. And 3 months after that, she began to pair words in "adjective-noun" and "noun-noun" pairs, such as "yellow car" and "Genie purse" (Fromkin et al., 1974). She was very focused on labels and had a much larger range of them than would be expected of a child at her linguistic stage (Curtiss, 1977). Eventually, she began to construct two-word sentences. She appeared to understand "wh-" questions ("who," "what," "when," "where," "why") but never really used them in her own communication, and it took her a very long time to understand and use personal pronouns like "I" and "me" (Curtiss et al., 1974; Fromkin et al., 1974).

Genie's linguistic abilities progressed far more slowly than her cognitive ones. When she was first found, she tested at the cognitive level of a 15-month-old; however, after 6 months of working with therapists and psychologists, she was able to think at the level of a 6- or 8-year-old (Fromkin et al., 1974). This speed of progress is astonishing, and it is very good support for the distinction between capability and ability. Genie's brain was clearly capable of developing certain cognitive functions, even though she had never been given the opportunity to actually learn to do them until she was discovered. Genie's case also illustrates that while the human brain has the amazing capacity to reorganize itself, some abilities, such as human language, may not develop as easily. This, then, provides evidence of a critical period in human language acquisition.

Psychologists and linguists worked with and studied Genie for approximately 7 years after she was found, at which point turmoil over her legal custody moved her several times between various foster homes and her biological mother. We know that Genie regressed to the point of complete silence at some point, like when she was living at home as a child. Currently, Genie lives in a care home in the Los Angeles area as a ward of the state. Beyond that, little is known. Most of the scientists she was close to have not been allowed to see her since 1977, let alone continue to monitor and investigate her language abilities (Grimshaw, Adelstein, Bryden, & MacKinnon, 1998; Jones, 1995; Rymer, 1993).

YOUR TURN!

Thanks to the Internet, the general public and researchers alike can share recordings of wild animals communicating with each other. In what are called playback studies, researchers in the field and laboratory play recordings of animals' mating calls, territory defense calls, alarm calls, or even the vocalizations of predators and observe their focal animals' behavior. This is how we learned, for example, about Diana monkeys' distinct leopard, snake, and eagle vocalizations.

Traditionally, field researchers lug heavy equipment into an animal's natural habitat in order to conduct playback studies. In this ministudy, we'll keep it simple. A smartphone (external speaker optional) and familiarity with ethogram building (Chapter 2, Theoretical and Methodological Approaches to Animal Cognition) are all you'll need. Keep in mind the research idea we'll share will be specific to birds, but we encourage you to get creative with other species and make use of the animal-audio-clip bounty that the internet has to offer.

Find a quiet place where you'll have good visual access to listen and observe a particular species. For avid bird watchers, this might be a favorite park or nature trail. Bird feeders are also a great place to do this (although remember that appropriate food should always be used in feeders). Identify the bird species you will likely be working with. There are many smartphone apps and websites to help with this. We recommend Cornell University's www.allaboutbirds.org database. Once you've identified your species, read as much as you can about that species, and come up with a research plan. If your focal

CHAPTER 4   COMMUNICATION BETWEEN ANIMALS

species is in its breeding season, you might find an audio clip of a mating call. On the other hand, if it's out of breeding season, playing a mating call might be equally, if not more, interesting. If there's a particular bird-of-prey species that is common in your area, you might get an audio clip of that species cued up as well. Decide on two or three different audio clips, with a rationale for your choice behind each.

Start by constructing a 3-minute baseline ethogram for your species without any playback audio. Maybe your baseline indicates that the bird pecks and eats four times, ruffles its feathers once, and vocalizes twice. If the bird is singing or making other kinds of vocalizations, try to characterize them as best you can (e.g., swooping, warbling, "three caws in a row," "chick-a-dee-ee-ee"). Once your ethogram has documented the bird's natural behavioral and vocal routine, introduce your different audio clips one at a time, and start observing the moment you hit play on the audio clip. Carefully listen, watch, and record changes by tallying behaviors from your original ethogram as well as any new ones that emerge. Binoculars might be helpful here for subtle behaviors like feather puffing or tail flicking. As with all field research, there may be some—or a lot—of waiting involved, especially if one of your audio clips causes the birds to high-tail it in anticipation of a predator. Just be patient and wait for your focal species to return, and then replay the clip and begin observing again. Or, if you're able, you could even follow the bird to see where it goes in response to your playback.

In addition to playing back different kinds of species-specific vocalizations or predator vocalizations, you might also do some observation of what happens when a playback of a completely nonnative bird is played. This would be the equivalent of showing up to your favorite restaurant and hearing a language you've never heard before. One research question might be to ask whether such a playback would cause a bird to abandon its territory or stand its ground, despite not knowing anything about the intruder except its song.

The playback technique's versatile methodology lends itself nicely to being used by citizen scientists. As mentioned, while this ministudy is specific to birds, we encourage you to adapt it for other species or pets, keeping in mind that many animals' main communication systems are olfactory or visual, rather than vocal. How might knowing this impact the results of your study? Collect some data, and find out!

REFERENCES

Ball, G. F., & Hulse, S. H. (1998). Birdsong. *American Psychologist, 55*(1), 37–58. doi:10.1037%2F0003-066X.53.1.37

Barón Birchenall, L. (2016). Animal Communication and Human Language: An overview. *International Journal of Comparative Psychology, 29*(1), Article ID 28000.

Barrett, L., & Würsig, B. (2014). Why dolphins are not aquatic apes. *Animal Behavior and Cognition, 1*(1), 1–18. doi:10.12966/abc.02.01.2014

Bee, M. A., Perrill, S. A., & Owen, P. C. (2000). Male green frogs lower the pitch of acoustic signals in defense of territories: A possible dishonest signal of size? *Behavioral Ecology, 11*(2), 169–177. doi:10.1093/beheco/11.2.169

Beran, M. J., Parrish, A. E., Perdue, B. M., & Washburn, D. A. (2014). Comparative cognition: Past, present, and future. *International Journal of Comparative Psychology, 27*(1), 3–30.

Boughman, J. W. (1998). Vocal learning by greater spear-nosed bats. *Proceedings in the Royal Society of London-Series B: Biological Sciences, 265*, 227–233. doi:10.1098/rspb.1998.0286

Caldwell, M. C., & Caldwell, D. K. (1965). Individualized whistle contours in bottle-nosed Dolphins (*Tursiops truncatus*). *Nature, 207*, 434–435. doi:10.1038/207434a0

Cartmill, E. A., & Byrne, R. W. (2007). Orangutans modify their gestural signaling according to their audience's comprehension. *Current Biology, 17*(15), 1345–1348. doi:10.1016/j.cub.2007.06.069

Chisholm, A. H., & Pearse, T. (1936). Injury feigning in birds. *The Auk, 53*(2), 251–253. doi:10.2307/4077341

Clegg, I. L. K., Rödel, H. G., Boivin, X., & Delfour, F. (2018). Looking forward to interacting with their caretakers: Dolphins' anticipatory behaviour indicates motivation to participate in specific events. *Applied Animal Behaviour Science, 202*, 85–93. doi:10.1016/j.applanim.2018.01.015

Colbert-White, E. N., Covington, M. A., & Fragaszy, D. M. (2011). Social context influences the vocalizations of a home-raised African grey parrot (*Psittacus erithacus erithacus*). *Journal of Comparative Psychology, 125*(2), 175–184. doi:10.1037/a0022097

Colbert-White, E. N., Hall, H. C., & Fragaszy, D. M. (2016). Variations in an African Grey parrot's speech patterns following ignored and denied requests. *Animal Cognition, 19*, 459–469. doi:10.1007/s10071-015-0946-1

Coye, C., Ouattara, K., Arlet, M. E., Lemasson, A., & Zuberbühler, K. (2018). Flexible use of simple and combined calls in female Campbell's monkeys. *Animal Behaviour, 141*, 171–181. doi:10.1016/J.ANBEHAV.2018.05.014

Curtiss, S. (1977). *Genie: A psycholinguistic study of a modern-day "wild child."* New York, NY: Academic Press.

Curtiss, S., Fromkin, V., Krashen, S. D., Rigler, D., & Rigler, M. (1974). The linguistic development of Genie. *Language, 50*(3), 528–554. doi:10.2307/412222

Czaczkes, T. J., Grüter, C., & Ratnieks, F. L. W. (2015). Trail Pheromones: An integrative view of their role in social insect colony organization. *Annual Review of Entomology, 60*(1), 581–599. doi:10.1146/annurev-ento-010814-020627

Deloache, J. S., Chiong, C., Sherman, K., Islam, N., Vanderborght, M., Troseth, G. L., … O'Doherty, K. (2010). Do babies learn from baby media? *Psychological Science : A Journal of the American Psychological Society/APS, 21*(11), 1570. doi:10.1177/0956797610384145

Díaz López, B., & Shirai, J. A. B. (2009). Mediterranean common bottlenose dolphin's repertoire and communication use. In A. G. Pearce & L. M. Correa (Eds.), *Dolphins: Anatomy, behavior, and threats* (pp. 129–148). New York, NY: Nova Science Publishers, Inc.

Doupe, A. J., & Kuhl, P. K. (1999). Birdsong and human speech: Common themes and mechanisms. *Annual Review of Neuroscience, 22*, 567–631. doi:10.1146/annurev.neuro.22.1.567

Duranton, C., Bedossa, T., & Gaunet, F. (2016). When facing an unfamiliar person, pet dogs present social referencing based on their owners' direction of movement alone. *Animal Behaviour, 113*, 147–156. doi:10.1016/j.anbehav.2016.01.004

Eimas, P. (1985). The perception of speech in early infancy. *Scientific American, 252*(1), 46–53. Retrieved from http://www.jstor.org/stable/24967546

Eisenberg, J. F., & Kleiman, D. G. (1972). Olfactory communication in mammals. *Source: Annual Review of Ecology and Systematics, 3*(137), 1–32. doi:10.1146/annurev.es.03.110172.000245

Fitch, W. T. (2000). The evolution of speech: A comparative review. *Trends in Cognitive Sciences, 4*(7), 258–267. doi:10.1016/S1364-6613(00)01494-7

Fitch, W. T., de Boer, B., Mathur, N., & Ghazanfar, A. A. (2016). Monkey vocal tracts are speech-ready. *Science Advances, 2*(12), e1600723. doi:10.1126/sciadv.1600723

Fitch, W. T., & Reby, D. (2001). The descended larynx is not uniquely human. *Proceedings of the Royal Society of London Series B-Biological Sciences, 268*(1477), 1669–1675. doi:10.1098/rspb.2001.1704

Flower, T. (2011). Fork-tailed drongos use deceptive mimicked alarm calls to steal food. *Proceedings. Biological Sciences, 278*(1711), 1548–1555. doi:10.1098/rspb.2010.1932

Flower, T. P., Gribble, M., & Ridley, A. R. (2014). Deception by flexible alarm mimicry in an African bird. *Science, 344*, 513–516. doi:10.1126/science.1249723

Fouts, R. S. (1974). Language: Origins, definition, and chimpanzees. *Journal of Human Evolution, 3*, 475–482. doi:10.1016/0047-2484(74)90007-4

Fouts, R. S. (1975). Capacities for language in great apes. In R. H. Tuttle (Ed.), *Society and Psychology of Primates* (pp. 371–390). The Hague: Mouton.

Fromkin, V., Krashen, S. D., Curtiss, S., Rigler, D., & Rigler, M. (1974). The development of language in Genie: A case of language acquisition beyond the critical period. *Brain and Language, 1*(1), 81–107. doi:10.1016/0093-934X(74)90027-3

Gardner, R. A., & Gardner, B. T. (1969). Teaching sign language to a chimpanzee. *Science, 165*(894), 664–672. doi:10.1126/science.165.3894.664

Garland, E. C., Goldizen, A. W., Rekdahl, M. L., Constantine, R., Garrigue, C., Hauser, N. D., … Noad, M. J. (2011). Dynamic horizontal cultural transmission of humpback whale song at the ocean basin scale. *Current Biology : CB, 21*(8), 687–691. doi:10.1016/j.cub.2011.03.019

Giret, N., Miklósi, Á., Kreutzer, M., & Bovet, D. (2009). Use of experimenter-given cues by African gray parrots (*Psittacus erithacus*). *Animal Cognition, 12*, 1–10. doi:10.1007/s10071-008-0163-2

Gould, J. L., Henerey, M., & MacLeod, M. C. (1970). Communication of Direction by the Honey Bee. *Science, 169*(3945), 544–554. Retrieved from http://www.ncbi.nlm.nih.gov/pubmed/5426774

Grimshaw, G. M., Adelstein, A., Bryden, M. P., & MacKinnon, G. E. (1998). First-language acquisition in adolescence: Evidence for a critical period for verbal language development. *Brain and Language, 63*(2), 237–255. doi:10.1006/brln.1997.1943

Grüter, C., Balbuena, M. S., & Farina, W. M. (2008). Informational conflicts created by the waggle dance. *Proceedings. Biological Sciences, 275*(1640), 1321–1327. doi:10.1098/rspb.2008.0186

Gyger, M., & Marler, P. (1988). Food calling in the domestic fowl, Gallus gallus: The role of external referents and deception. *Animal Behaviour, 36*(2), 358–365. doi:10.1016/S0003-3472(88)80006-X

Handel, S., Todd, S. K., Zoidis, A. M., Handel, S., Todd, S. K., & Hierarchical, A. M. Z. (2012). Hierarchical and rhythmic organization in the songs of humpback whales (*Megaptera novaeangliae*). *Bioacoustics, 12*(2), 141–156. doi:10.1080/09524622.2012.668324

Hardus, M. E., Lameira, A. R., Van Schaik, C. P., & Wich, S. A. (2009). Tool use in wild orang-utans modifies sound production: A functionally deceptive innovation? *Proceedings in the Royal Society of London-Series B: Biological Sciences, 276*(1673), 3689–3694. doi:10.1098/rspb.2009.1027

Hare, B. A. (2017). Survival of the friendliest: Homo sapiens evolved via selection for prosociality. *Annual Review of Psychology, 68*, 155–186. doi:10.1146/annurev-psych-010416-044201

Hare, B. A., Brown, M., Williamson, C., & Tomasello, M. (2002). The domestication of social cognition in dogs. *Science, 298*, 1634–1636. doi:10.1126/science.1072702

Hare, B. A., Plyusnina, I., Ignacio, N., Schepina, O., Stepika, A., Wrangham, R., & Trut, L. N. (2005). Social cognitive evolution in captive foxes is a correlated by-product of experimental domestication. *Current Biology, 15*, 226–230. doi:10.1016/j.cub.2005.01.040

Herman, L. M. (2010). What laboratory research has told us about dolphin cognition. *International Journal of Comparative Psychology, 23*(3), 310–330. Retrieved from https://escholarship.org/uc/uclapsych_ijcp

Herman, L. M., Abichandani, S. L., Elhajj, A. N., Herman, E. Y. K., Sanchez, J. L., & Pack, A. A. (1999). Dolphins (*Tursiops truncatus*) comprehend the referential character of the human pointing gesture. *Journal of Comparative Psychology, 113*(4), 347–364. doi:10.1037/0735-7036.113.4.347

Herman, L. M., & Forestell, P. H. (1985). Reporting presence or absence of named objects by a language-trained dolphin. *Neuroscience & Biobehavioral Reviews, 9*(4), 667–681. doi:10.1016/0149-7634(85)90013-2

Herman, L. M., Kuczaj, S. A. II., & Holder, M. D. (1993). Responses to anomalous gestural sequences by a language-trained dolphin: Evidence for processing of semantic relations and syntactic information. *Journal of Experimental Psychology-General, 122*(2), 184–194.

Herman, L. M., Morrel-Samuels, P., & Pack, A. A. (1990). Bottlenosed dolphin and human recognition of veridical and degraded video displays of an artificial gestural language. *Journal of Experimental Psychology: General, 119*, 215–230. doi:10.1037/0096-3445.119.2.215

Herman, L. M., Richards, D. G., & Wolz, J. P. (1984). Comprehension of sentences by bottlenosed dolphins. *Cognition, 16*(2), 129–219. doi:10.1016/0010-0277(84)90003-9

Herzing, D. L. (1996). Vocalizations and associated underwater behavior of free-ranging Atlantic spotted dolphins, *Stenella frontalis* and bottlenose dolphins, *Tursiops truncatus*. *Aquatic Mammals, 22*(2), 61–79. doi:10.12966/abc.02.02.2015

Hess, E. (2008). *Nim Chimpsky: The chimp who would be human*. New York, NY: Bantam Books.

Hile, A. G., Plummer, T. K., & Striedter, G. F. (2000). Male vocal imitation produces call convergence during pair bonding in budgerigars, *Melopsittacus undulatus*. *Animal Behaviour, 59*(6), 1209–1218. doi:10.1006/anbe.1999.1438

Hobaiter, C., & Byrne, R. W. (2014). The meanings of chimpanzee gestures. *Current Biology, 24*(14), 1596–1600. doi:10.1016/j.cub.2014.05.066

Hobaiter, C., & Byrne, R. W. (2017). What is a gesture? A meaning-based approach to defining gestural repertoires. *Neuroscience and Biobehavioral Reviews, 82*, 3–12. doi:10.1016/j.neubiorev.2017.03.008

Hostetter, A. B., Cantero, M., & Hopkins, W. D. (2001). Differential use of vocal and gestural communication by chimpanzees (*Pan troglodytes*) in response to the attentional status of a human (*Homo sapiens*). *Journal of Comparative Psychology, 115*(4), 337–343. doi:10.1037/0735-7036.115.4.337

Itakura, S., & Tanaka, M. (1998). Use of experimenter-given cues during object-choice tasks by chimpanzees (*Pan troglodytes*), an orangutan (*Pongo pygmaeus*) and human infants (*Homo sapiens*). *Journal of Comparative Psychology, 112*, 119–126. doi:10.1037/0735-7036.112.2.119

Jackson, K., & Jackson, B. (1950). *Little yip-yip and his bark*. New York, NY: Simon and Schuster.

Janik, V. M., & Slater, P. (1998). Context-specific use suggests that bottlenose dolphin signature whistles are cohesion calls. *Animal Behavior, 56*, 829–838. doi:10.1006/anbe.1998.0881

Johnson, J. S., & Newport, E. L. (1989). Critical period effects in second language learning: The influence of maturational state on the acquisition of English as a second language. *Cognitive Psychology, 21*(1), 60–99. doi:10.1016/0010-0285(89)90003-0

Jones, P. E. (1995). Contradictions and unanswered questions in the Genie case: A fresh look at the linguistic evidence. *Language & Communication, 15*(3), 261–280. doi:10.1016/0271-5309(95)00007-D

Kako, E. (1999). Elements of syntax in the systems of three language-trained animals. *Animal Learning & Behavior, 27*(1), 1–14. doi:10.3758/BF03199424

Kaufman, A. B., Colbert-White, E. N., & Burgess, C. (2013). Higher-order semantic structures in an African Grey parrot's vocalizations: Evidence from the hyperspace analog to language (HAL) model. *Animal Cognition, 16*(5), 789–801. doi:10.1007/s10071-013-0613-3

Kaufman, A. B., Green, S. R., Seitz, A. R., & Burgess, C. (2012). Using a self-organizing map (SOM) and the Hyperspace Analog to Language (HAL) model to identify patterns of syntax and structure in the songs of humpback whales. *International Journal of Comparative Psychology, 35*, 237–270.

King, S. L., Harley, H. E., & Janik, V. M. (2014). The role of signature whistle matching in bottlenose dolphins, *Tursiops truncatus*. *Animal Behaviour, 96*, 79–86. doi:10.1016/j.anbehav.2014.07.019

Kuhl, P. K. (2004). Early language acquisition: Cracking the speech code. *Nature Reviews Neuroscience, 5*(11), 831–843. doi:10.1038/nrn1533

Kuhl, P. K., Conboy, B. T., Coffey-Corina, S., Padden, D., Rivera-Gaxiola, M., & Nelson, T. (2008). Phonetic learning as a pathway to language: New data and native language magnet theory expanded (NLM-e). *Philosophical Transactions of the Royal Society B-Biological Sciences, 363*(1493), 979–1000. doi:10.1098/rstb.2007.2154

Lenneberg, E. H. (1967). *Biological foundations of language*. Oxford, UK: Wiley.

Liebal, K., Pika, S., Call, J., & Tomasello, M. (2004). To move or not to move: How apes adjust to the attentional state of others. *Interaction Studies, 5*, 199–219. doi:10.1075/is.5.2.03lie

Lloyd, J. E. (1984). Occurrence of aggressive mimicry in fireflies. *Florida Entomologist, 67*, 368–376. doi:10.2307/3494715

López, B. D. (2011). Whistle characteristics in free-ranging bottlenose dolphins (*Tursiops truncatus*) in the Mediterranean Sea: Influence of behaviour. *Mammalian Biology, 76*, 180–189. doi:10.1016/j.mambio.2010.06.006

Madsen, P. T., Jensen, F. H., Carder, D., & Ridgway, S. (2011). Dolphin whistles: A functional misnomer revealed by heliox breathing. *Biology letters, 8*(2), 211–213. doi:10.1098/rsbl.2011.0701

Maros, K., Gácsi, M., & Miklósi, Á. (2008). Comprehension of human pointing gestures in horses (*Equus caballus*). *Animal Cognition, 11*, 457–466. doi:10.1007/s10071-008-0136-5

McCowan, B., & Reiss, D. (1997). Quantitative comparison of whistle repertoires from captive adult bottle-nosed dolphins (*Delphinidae, Tursiops-Truncatus*): a reevaluation of the signature whistle hypothesis. *Ethology, 100*(3), 194–209. doi:10.1111/j.1439-0310.1995.tb00325.x

Miklósi, A., Kubinyi, E., Topál, J., Viranyi, Z., & Csányi, V. (2003). A simple reason for a big difference: Wolves do not look back at humans, but dogs do. *Current Biology, 13*, 763–766. doi:10.1016/s0960-9822(03)00263-x

Miklósi, Á., Pongrácz, P., Lakatos, G., Topál, J., & Csányi, V. (2005). A comparative study of the use of visual communicative signals in interactions between dogs (*Canis familiaris*) and humans and cats (*Felis catus*) and humans. *Journal of Comparative Psychology, 119*(2), 179. doi:10.1037/0735-7036.119.2.179

Møller, A. P. (1988). False alarm calls as a means of resource usurption in the Great Tit Parus major. *Ethology, 79*(1), 25–30. doi:10.1111/j.1439-0310.1988.tb00697.x

Morford, J. P., Grieve-Smith, A. B., MacFarlane, J., Staley, J., & Waters, G. (2008). Effects of language experience on the perception of American Sign Language. *Cognition, 109*(1), 41–53. doi:10.1016/j.cognition.2008.07.016

Morisaka, T., Shinohara, M., Nakahara, F., & Akamatsu, T. (2005). Geographic variations in the whistles among three Indo-Pacific bottlenose dolphin Tursiops aduncus populations in Japan. *Fisheries Science, 71*(3), 568–576. doi:10.1111/j.1444-2906.2005.01001.x

Nawroth, C., & von Borell, E. (2015). Domestic pigs'(*Sus scrofa domestica*) use of direct and indirect visual and auditory cues in an object choice task. *Animal Cognition, 18*(3), 757–766. doi:10.1007%2Fs10071-015-0842-8

Newport, E. L. (1990). Maturational constraints on language learning. *Cognitive Science, 14*, 11–28. doi:10.1207/s15516709cog1401

Norman, M. D., Finn, J., & Tregenza, T. (1999). Female impersonation as an alternative reproductive strategy in giant cuttlefish. *Proceedings of the Royal Society of London B: Biological Sciences, 266*(1426), 1347–1349. doi:10.1098/rspb.1999.0786

Nottebohm, F. (1972). The origins of vocal learning. *The American Naturalist, 106*(947), 116–140. doi:10.1086/282756

Ouattara, K., Lemasson, A., & Zuberbühler, K. (2009). Campbell's monkeys concatenate vocalizations into context-specific call sequences. *Proceedings of the National Academy of Sciences of the United States of America, 106*(51), 22026–22031. doi:10.1073/pnas.0908118106

Ouattara, K., Zuberbühler, K., N'goran, E. K., Gombert, J.-E., & Lemasson, A. (2009). The alarm call system of female Campbell's monkeys. *Animal Behaviour, 78*(1), 35–44. doi:10.1016/j.anbehav.2009.03.014

Pack, A. A., & Herman, L. M. (2007). The dolphin's (*Tursiops truncatus*) understanding of human gazing and pointing: Knowing what and where. *Journal of Comparative Psychology, 121*(1), 34–45. doi:10.1037/0735-7036.121.1.34

Papale, E., Perez-Gil, M., Castrillon, J., Perez-Gil, E., Ruiz, L., Servidio, A., ... Martín, V. (2017). Context specificity of Atlantic spotted dolphin acoustic signals in the Canary Islands. *Ethology Ecology & Evolution, 29*(4), 311–329. doi:10.1080/03949370.2016.1171256

Patterson, F. G. (1986). The mind of the gorilla: Conversation and conservation. In K. Benirschke (Ed.), *Primates: The road to selfsustaining populations* (pp. 933–947). New York, NY: Springer-Verlag.

Patterson, F. G., & Linden, E. (1981). *The education of Koko*. New York, NY: Holt, Rinehart and Winston.

Payne, K. (2003). Sources of social complexity in the three elephant species. In F. B. M. de Waal & P. J. Tyack (Eds.), *Animal social complexity: Intelligence, culture, and individualized societies* (pp. 261–287). Cambridge, MA: Harvard University Press.

Penfield, W., & Roberts, L. (1959). *Speech and brain mechanisms*. Princeton, NJ: Princeton University Press.

Pepperberg, I. M. (1981). Functional vocalizations by an African Grey parrot (*Psittacus erithacus*). *Zeitschrift Für Tierpsychologie, 55*(2), 139–160. Retrieved from http://onlinelibrary.wiley.com/doi/10.1111/j.1439-0310.1981.tb01265.x/abstract

Pepperberg, I. M. (1992). Proficient performance of a conjunctive, recursive task by an African Grey parrot (*Psittacus erithacus*). *Journal of Comparative Psychology, 106*(3), 295–305. doi:10.1037/0735-7036.106.3.295

Pepperberg, I. M. (1999). *The Alex studies: Cognitive and communicative abilities of grey parrots*. Cambridge, MA: Harvard University Press.

Pepperberg, I. M. (2006). Cognitive and communicative abilities of Grey parrots. *Applied Animal Behaviour Science, 100*(1–2), 77–86. doi:10.1016/j.applanim.2006.04.005

Pepperberg, I. M. (2012). Further evidence for addition and numerical competence by a Grey parrot (*Psittacus erithacus*). *Animal Cognition, 15*(4), 711–717. doi:10.1007/s10071-012-0470-5

Pepperberg, I. M. (2013). Abstract concepts: Data from a Grey parrot. *Behavioural Processes, 93*, 82–90. doi:10.1016/j.beproc.2012.09.016

Pepperberg, I. M., Brese, K. J., & Harris, B. J. (1991). Solitary sound play during acquisition of English vocalizations by an African Grey parrot (*Psittacus erithacus*): Possible parallels with children's monologue speech. *Applied Psycholinguistics, 12*(02), 151–178. Retrieved from http://journals.cambridge.org/abstract_S0142716400009127

Pepperberg, I. M., & Gordon, J. D. (2005). Number comprehension by a grey parrot (*Psittacus erithacus*), including a zero-like concept. *Journal of Comparative Psychology, 119*(2), 197–209. doi:10.1037/0735-7036.119.2.197

Pilley, J. W., & Reid, A. K. (2011). Border collie comprehends object names as verbal referents. *Behavioural Processes, 86*(2), 184–195. doi:10.1016/j.beproc.2010.11.007

Plotnik, J. M., & de Waal, F. B. (2014). Asian elephants (*Elephas maximus*) reassure others in distress. *PeerJ, 2*, e278. doi:10.7717/peerj.278

Poole, J. H., Tyack, P. L., Stoeger-Horwath, A. S., & Watwood, S. (2005). Animal behaviour: Elephants are capable of vocal learning. *Nature, 434*(7032), 455. doi:10.1038/434455a

Preece, K., & Beekman, M. (2014). Honeybee waggle dance error: Adaption or constraint? Unravelling the complex dance language of honeybees. *Animal Behaviour, 94*, 19–36. doi:10.1016/j.anbehav.2014.05.016

Quick, N. J., & Janik, V. M. (2012). Bottlenose dolphins exchange signature whistles when meeting at sea. *Proceedings. Biological Sciences/The Royal Society, 279*(1738), 2539–2545. doi:10.1098/rspb.2011.2537

Rehn, N., Filatova, O. A., Durban, J. W., & Foote, A. D. (2010). Cross-cultural and cross-ecotype production of a killer whale 'excitement' call suggests universality. *Naturwissenschaften, 98*(1), 1–6. doi:10.1007/s00114-010-0732-5

Richards, D. G., Wolz, J. P., & Herman, L. M. (1984). Vocal mimicry of computer generated sounds and vocal labeling of objects by a bottlenosed dolphin, *Tursiops truncatus. Journal of Comparative Psychology, 98*(1), 10–28. doi:10.1037/0735-7036.98.1.10

Roberts, A. I., Vick, S.-J., & Buchanan-Smith, H. M. (2012). Usage and comprehension of manual gestures in wild chimpanzees. *Animal Behaviour, 84*(2), 459–470. Retrieved from http://www.sciencedirect.com/science/article/pii/S0003347212002424

Rumbaugh, D. M., & Gill, T. V. (1977). Lana's acquisition of language skills. In D. M. Rumbaugh (Ed.), *Language learning by a chimpanzee: The Lana project* (pp. 165–192). New York, NY: Academic Press.

Rymer, R. (1993). *Genie: Escape from a Silent Childhood*. London, UK: Michael Joseph.

Sanvito, S., Galimberti, F., & Miller, E. H. (2007). Observational evidences of vocal learning in southern elephant seals: A longitudinal study. *Ethology, 113*, 137–146. doi:10.1111/j.1439-0310.2006.01306.x

Savage-Rumbaugh, E. S. (1986). *Ape language: From conditioned response to symbol*. New York, NY: Columbia University Press.

Savage-Rumbaugh, E. S., & Lewin, R. (1994). *Kanzi: The ape at the brink of the human mind*. New York, NY: Wiley and Sons.

Savage-Rumbaugh, E. S., Rumbaugh, D. M., & Boysen, S. T. (1978). Symbolic communication between two chimpanzees (*Pan troglodytes*). *Science, 201*(4356), 641. Retrieved from http://www.sciencemag.org/content/201/4356/641.short

Savage-Rumbaugh, E. S., Rumbaugh, D. M., & Fields, W. M. (2009). Empirical Kanzi: The ape language controversy revisited. *Skeptic, 15*(1), 25–33.

Sayigh, L. S. (1999). Individual recognition in wild bottlenose dolphins: A field test using playback experiments. *Animal Behaviour, 57*(1), 41–50. doi:10.1006/anbe.1998.0961

Sayigh, L. S., Esch, H. C., Wells, R. S., & Janik, V. M. (2007). Facts about signature whistles of bottlenose dolphins, *Tursiops truncatus*. *Animal Behaviour, 74*(6), 1631–1642. doi:10.1016/j.anbehav.2007.02.018

Schel, A. M., Machanda, Z., Townsend, S. W., Zuberbühler, K., & Slocombe, K. E. (2013). Chimpanzee food calls are directed at specific individuals. *Animal Behaviour, 86*(5), 955–965. Retrieved from http://www.sciencedirect.com/science/article/pii/S0003347213003813

Schel, A. M., Townsend, S. W., Machanda, Z., Zuberbühler, K., & Slocombe, K. E. (2013). Chimpanzee alarm call production meets key criteria for intentionality. *PLoS One, 8*(10), e76674. doi:10.1371/journal.pone.0076674

Schel, A. M., Tranquilli, S., & Zuberbühler, K. (2009). The alarm call system of two species of black-and-white colobus monkeys (*Colobus polykomos* and *Colobus guereza*). *Journal of Comparative Psychology, 123*(2), 136–150. doi:10.1037/a0014280

Scheumann, M., & Call, J. (2004). The use of experimenter-given cues by South African fur seals (*Arctocephalus pusillus*). *Animal Cognition, 7*, 224–230. doi:10.1007/s10071-004-0216-0

Sevcik, R. A., & Savage-Rumbaugh, E. S. (1994). Language comprehension and use by great apes. *Language and Communication, 14*(1), 37–58. doi:10.1016/0271-5309(94)90019-1

Sewall, K. B., Young, A. M., & Wright, T. F. (2016). Social calls provide novel insights into the evolution of vocal learning. *Animal Behaviour, 120*, 163–172. doi:10.1016/j.anbehav.2016.07.031

Seyfarth, R. M., & Cheney, D. L. (2003). Signalers and receivers in animal communication. *Annual Review of Psychology, 54*, 145–173. doi:10.1146/annurev.psych.54.101601.145121

Seyfarth, R. M., & Cheney, D. L. (2017). The origin of meaning in animal signals. *Animal Behaviour, 124*, 339–346. doi:10.1016/j.anbehav.2016.05.020

Seyfarth, R. M., Cheney, D. L., & Marler, P. (1980). Vervet monkey alarm calls: Semantic communication in a free-ranging primate. *Animal Behavior, 28*, 1070–1094. doi:10.1016/S0003-3472(80)80097-2

Shanker, S. G., Savage-Rumbaugh, E. S., & Taylor, T. J. (1999). Kanzi: A new beginning. *Animal Learning and Behavior, 27*(1), 24–25. doi:10.3758/BF03199427

Shettleworth, S. J. (2010). Clever animals and killjoy explanations in comparative psychology. *Trends in Cognitive Sciences, 14*(11), 477–481. doi:10.1016/j.tics.2010.07.002

Slobodchikoff, C. N. (2002). Cognition and communication in prairie dogs. In M. Beckoff, C. Allen, & G. M. Burghardt (Eds.), (pp. 217–228). Cambridge, MA: MIT.

Slobodchikoff, C. N., Kiriazis, J., Fischer, C., & Creef, E. (1991). Semantic information distinguishing individual predators in the alarm calls of Gunison's prairie dogs. *Animal Behavior, 42*, 713–719. doi:10.1016/S0003-3472(05)80117-4

Suzuki, T. N., Wheatcroft, D., & Griesser, M. (2016). Experimental evidence for compositional syntax in bird calls. *Nature Communications, 7*, 1–7. doi:10.1038/ncomms10986

Terrace, H. S., Petitto, L.-A., Sanders, R. J., & Bever, T. G. (1979). Can an ape create a sentence? *Science, 206*(4421), 891–902. doi:10.1126/science.504995

Thienen, W. Von, Metzler, D., Choe, D. H., & Witte, V. (2014). Pheromone communication in ants: A detailed analysis of concentration-dependent decisions in three species. *Behavioral Ecology and Sociobiology, 68*(10), 1611–1627. doi:10.1007/s00265-014-1770-3

Thiessen, E. D., & Saffran, J. R. (1993). When cues collide: Use of stress and statistical cues to word boundaries by 7- to 9-month-old infants. *Developmental Psychology, 39*(4), 706–716. doi:10.1037/0012-1649.39.4.706

Trut, L. N. (1999). Early canid domestication: The farm-fox experiment. *American Scientist, 87*(2), 160–169. doi:10.1511/1999.2.160

Viranyi, Z., Gacsi, M., Kubinyi, E., Topál, J., Belenyi, B., Ujfalussy, D., & Miklósi, A. (2008). Comprehension of human pointing gestures in young human-reared wolves (*Canis lupus*) and dogs (*Canis familiaris*). *Animal Cognition, 11*, 373–387. doi:10.1007/s10071-007-0127-y

von Frisch, K. (1967). *The dance language and orientation of bees*. Cambridge, MA: Harvard University Press.

Werker, J. F., & Tees, R. C. (1984). Cross-language speech perception: Evidence for perceptual reorganization during the first year of life. *Infant Behavior & Development, 7*(1), 49–63. doi:10.1016/S0163-6383(84)80022-3

Wheeler, B. C. (2009). Monkeys crying wolf? Tufted capuchin monkeys use anti-predator calls to usurp resources from conspecifics. *Proceedings of the Royal Society B: Biological Sciences, 276*(1669), 3013–3018. doi:10.1098/rspb.2009.0544

Wilson, E. O. (1958). The beginnings of nomadic and group-predatory behavior in the ponerine ants. *Evolution, 12*(1), 24–31. doi:10.2307/2405901

Winn, H. E., Thompson, T. J. J., Cummings, W. C., Hain, J., Hudnall, J., Hays, H., & Steiner, W. W. W. (1981). Song of the humpback whale—Population comparisons. *Behavioral Ecology and Sociobiology, 8*, 41–46. doi:10.1007/BF00302842

Winn, H. E., & Winn, L. K. (1978). The Song of the Humpback Whale in the West Indies. *Marine Biology, 47*, 97–114. doi:10.1007/BF00395631

Xitco Jr., M. J., Gory, J. D., & Kuczaj, S. A. II. (2001). Spontaneous pointing by bottlenose dolphins (*Tursiops truncatus*). *Animal Cognition, 4*, 115–123. doi:10.1007/s100710100107

Xitco Jr., M. J., Gory, J. D., & Kuczaj, S. A. II. (2004). Dolphin pointing is linked to the attentional behavior of a receiver. *Animal Cognition, 7*, 231–238. doi:10.1007/s10071-004-0217-z

Zuberbühler, K., Cheney, D. L., & Seyfarth, R. M. (1999). Conceptual semantics in a nonhuman primate. *Journal of Comparative Psychology, 113*(1), 33–42. doi:10.1037/0735-7036.113.1.33

Social Cognition in Animals

Within the animal kingdom, sociality is on a giant continuum, with a large degree of diversity in how social, with whom, and how complex those interactions are among conspecifics. We consider some species to be solitary, where adults may only interact when it's time to mate or defend territory. On the other end of the spectrum are highly social species, with complex social behaviors like division of labor. Division of labor involves having specialized roles for each individual, which allows them to cooperate with and support the rest of the group. Meerkats (*Suricata suricatta*; Clutton-Brock, Russell, & Sharpe, 2004), ants (*Pogonomyrmex californicus*; Holbrook, Barden, & Fewell, 2011), and even some shrimp (*Synalpheus regalis*; Duffy, 2003) are just a few examples of species whose group members have distinct roles. Other highly social species have distinct "names," which allow individuals within a group to recognize each other and cultivate deeper social relationships. Individual naming has been observed in species as predictable as parrots (*Forpus passerinus*) and dolphins (Berg, Delgado, Cortopassi, Beissinger, & Bradbury, 2011; Caldwell & Caldwell, 1965) and as surprising as bats (*Phyllostomus hastatus*; Boughman & Wilkinson, 1998).

In this chapter, we explore in greater depth some of the advanced ways that animals engage with one another. As we'll see, there appears to be a correlation between sociality and cognition. Knowing something about the depth (or lack thereof) of a species' social behavior allows researchers to contextualize and better understand cognitive abilities such as theory of mind, problem solving, and referential signaling in communication.

LEARNING FROM OTHERS

Imagine a time when you went to a place that was completely new. This could be a new country with very different customs, traditions, and cuisine; a first-time trip to a yoga class at a new fitness center; or your first day of junior high, high school, or postsecondary school. How did you figure out what to do and how to act once you were there? For example, if it was a new country, you might need to know how to judge if you were receiving good or bad service at a restaurant. If it was yoga, you would need to know whether to wear your shoes or take them off. At a new school, you might listen to others to learn whether to call instructors Mrs., Dr., or by their first name.

What resources did you use to gather the information you needed about the new place you imagined? Chances are, you might have done a quick Internet search, asked someone who had been there before, or simply watched the behavior of others when you arrived. We humans are fantastic at soaking up information from conspecifics. By learning from others, we more effectively and efficiently interact with our environment. When you notice no one is rolling their eyes or grimacing at their server, you can deduce that it's normal to wait an hour for food to arrive in Country X because, in their culture, restaurants are a place to both eat *and* socialize. You can avoid marking up the floor at the fitness center by reading the yoga class rules online before you go, and so on.

One of the downfalls of the behaviorist movement was social psychologist Albert Bandura's (1925–) objection to B. F. Skinner's claim that all behavior could be explained by consequences in the environment. Bandura's (1976) social learning theory argued that operant and classical conditioning simply could not explain all behavior—we learn from watching others as well! When I (ECW) see pedestrians crossing and decide it must also be safe for me to step out into the road, my behavior has been affected by social partners, but no one has reinforced or punished me for what I learned from watching. Sure, reinforcement and punishment can be relevant, such as observing a classmate get chewed out for calling their instructor by her first name, but the punishment did not happen to the individual doing the observing. That is what makes learning from others so special: no trial and error, no punishment associated with error, no wasted energy associated with multiple ineffective attempts, just all the benefits associated with quick and easy learning.

According to the social intelligence hypothesis (see Byrne & Whiten, 1988), there is much to be gained from being a member of a social species. In fact, many scientists attribute the increase in brain size in social species to the need for increased capacity to deal with complex social relationships. In other words, the social intelligence hypothesis would predict that as our human ancestors' societies and cultures became more complex, so did our ability to understand complex societies and cultures—and vice versa. This includes greater protection, more opportunities to learn efficient and effective ways of doing things, and access to resources found by group members. Given all the perks of group living, it should come as no surprise that from cephalopods to mammals, and many animals in between, there is strong evidence that other species learn from and copy the actions of others. For example, apes can learn a cooperative task by observing others (Suchak, Watzek, Quarles, & de Waal, 2018), bumblebees (*Bombus terrestris*) rely on social cues they get from other bees to help them with foraging (Avarguès-Weber, Lachlan, & Chittka, 2018), rats learn from other rats' mistakes (Bem, Jura, Bontempi, & Meyrand, 2018), dogs can learn how to solve a puzzle by observing a model (Miller, Rayburn-Reeves, & Zentall, 2009), octopuses (*Octopus vulgaris*) can learn color discrimination (Fiorito & Scotto, 1992) and how to open a jar containing food by watching a conspecific (Fiorito, Rosciano, Scotto, & Valsecchi, 1994), and the list continues.

Learning from others comes in three flavors. By reading the rules on the fitness center's website or listening in on conversations between the college professor and student, we learn from others in a way that is less direct, less intentional on the part of the model, and therefore more inferential on the part of the observer. We refer to this as observational learning. On the other hand, when you read the emotional expressions of restaurant patrons in Country X, ask a model directly for assistance, or role-play with a model to learn new skills, we refer to these more intentional acts as social learning. There is a great deal of gray area, and some might argue that the terms *observational* and *social* could be used interchangeably. But, generally speaking, if the model's behavior is geared toward personally teaching the individual in a socially intimate manner, you are likely experiencing social learning.

The human psychology and education communities (as well as some animal researchers!) have been against referring to "teaching" when describing social learning in animals, mainly because teaching implies a level of intentionality on the part of the model that is difficult to measure in animals. Nonetheless, there has been a movement within the animal cognition community to say that animals can, in fact, teach one another (Hoppitt et al., 2008). For example, skilled ants engage in a behavior called *tandem running*, in which they touch their bodies to the body of a novice ant as they lay down chemical trails, presumably to assist the newcomer with route learning (*Temnothorax albipennis*; Franks & Richardson, 2006). Killer whales also repeat the same seal hunting technique in front of their offspring, sometimes without even killing the seal, leading researchers to ask why they would repeatedly catch and release a seal if they were not planning to eat it (Lopez & Lopez, 1985). Considering the amount of energy they'd have to expend, there would need to be a good reason, and that reason might be teaching.

A third form of learning from others that arises from observing the behavior of a model is copying. Some behaviors are easier to copy than others, with the difficulty depending upon three factors: cognitive complexity required to copy, amount of effort or energy required to copy, and the degree to which the copied action was voluntarily produced. There are four "levels" of copying. At the bottom of the hierarchy, the least demanding form of copying, is contagious behavior, actions that are elicited automatically by a social partner. Contagious behaviors are instinctive, reflexive, and unconsciously produced, such as when we involuntarily feed off one another's laughter or yawn when someone else yawns. (Interestingly, a tendency to contagiously yawn has been correlated with empathy [Platek, Critton, Myers, & Gallup, 2003], another topic covered in this chapter.)

Contagious behaviors have been highly documented in the rest of the animal kingdom, due in part to the mechanical simplicity of them. Some behaviors are under the control of an innate modal action pattern (MAP), a behavioral sequence that runs to completion once activated, kind of like tipping over the edge on a roller coaster. Give a capuchin monkey any strong-smelling food like an onion, and it will set to work rubbing itself down with it—possibly as an innate, parasite-deterrent behavior. Those in the group who are not holding onions will also begin rubbing themselves down, kind of like how seeing someone with a bug crawling on them will cause you to scratch, too. Fur rubbing is just one of many examples of contagious behavior in animals. Other examples of contagious behaviors include yawning, some courtship displays, canine play bowing, spider web spinning, and geese rolling egg-shaped objects toward their nests. The beauty of contagious behaviors is that by automating some behaviors, animals can devote more energy and cognition to more intentional behaviors.

Stimulus enhancement and local enhancement are a bit higher up on the hierarchy of copying. Here, a model's interaction with a particular object or location, respectively, increases a social partner's interaction with it, as when a young child chooses to play with the same toy because an older sibling just finished playing with it. Here, the sibling's actions with the toy enhanced its value compared to all the other toys in the toy chest.[1] Humans are not the only ones drawn to objects and locations preferred by conspecifics. To illustrate just how diverse these behaviors are among both highly social and solitary species, stimulus and/or local enhancement has been observed in primates (e.g., Fragaszy, Deputte, Cooper, Colbert-White, & Hémery, 2011), rats (Heyes, Ray, Mitchell, & Nokes, 2000), some birds (e.g., graylag goose, *Anser anser*; Fritz, Bisengerger, & Kotrschal, 2000; Eurasian jays, *Garrulus glandarius*, Miller, Logan, Lister, & Clayton, 2016), bumblebees (Avarguès-Weber & Chittka, 2014), dogs (Cracknell, Simon Mills, & Kaulfuss, 2008), frog tadpoles (*Lithobates sylvatica*) and salamander larvae (*Ambystoma maculatum*; Chapman, Holcomb, Spivey, Sehr, & Gall, 2015), turtles (*Pseudemys nelsoni*; Davis & Burghardt, 2011), fish (see Brown & Laland, 2003), and some noninsect invertebrates like pond snails (*Lymnaea stagnalis*; see Webster & Fiorito, 2001).

What is the evolutionary value of being reflexively drawn to objects and locations that a social partner interacted with? The possibility of finding food, water, mates, shelter, or safety, among other resources, is too good to pass up, so it pays to check it out for yourself. Thus, from larvae to adults, no matter what your species is or

---

[1]. Note of validation for any parents reading: Your child *does* always want what their sibling has.

how social you are, it pays to attend to what others are doing and where they're doing it.

Third on the hierarchy is emulation, which happens when an observer uses similar but different steps to achieve the same goal as a model. The observer does not need to precisely copy each step of the behavior, but they need to produce the same end result. Like the saying "There's more than one way to skin a cat," with emulation, it does not matter how you do it as long as you achieve the goal. A newcomer to the sport of volleyball might copy the end goal of making the ball bounce on the opponent's side of the net, but they might do so by hitting the ball differently from how the model demonstrated it to them during lessons.

Some say that emulation requires an individual to understand the cause and effect of the model's behavior. Horner and Whiten (2005) tested this by presenting chimpanzees and young children with a clear plastic box, which they tapped, poked, and manipulated in a series of steps before opening a chute to reveal a reward. Children copied the steps exactly, even though none of the steps contributed to making the reward appear in the chute. Looking through the clear plastic box made this obvious. In sharp contrast to the old "monkey see, monkey do," the apes skipped the tapping, sliding, and poking, and went straight to open the chute. That is, while both the children and apes observed the meaningless steps, the apes only copied the end-goal step of opening the chute. Sure, it might seem like the apes were "smarter" than the children for not being fooled, but a good comparative researcher would look to the animals' natural history for answers before making such a bold claim. Humans are highly sensitive to learning from other humans, especially when we are young. Society tells us that adults are smart and know everything, so why would we not copy everything one does? On the other hand, apes do not appear to place humans on such a social learning pedestal, which is why it's understandable that they filtered through the useless information and went straight for the reward.

While Horner and Whiten's (2005) apes engaged in emulation, the children used imitation, the highest level of copying. Imitation involves copying both the process and end goal of the behavior, such as perfecting a backflip or flawlessly replicating Whitney Houston's iconic voice. Because imitation requires greater attention, skill matching, and effort to replicate all steps of the behavior, imitation is considered highly complex. True imitation by animals is less common than contagious behavior, stimulus/local enhancement, or

CHAPTER 5 SOCIAL COGNITION IN ANIMALS

even emulation, but it has been observed—largely in primates and dolphins. For example, dolphins trained in the "do as I do" paradigm can imitate motor and vocal behaviors of another dolphin (Jaakkola, Guarino, & Rodriguez, 2010). What's even more remarkable is that they can copy these behaviors while blindfolded! Imagine being blindfolded and told "Do what your friend is doing," with no further assistance or explanation. Somehow, either via direct communication that is beyond the scope of our understanding, by feeling the water pressure patterns, or something else we have not considered, a blindfolded dolphin can imitate the complex acrobatic behavior its conspecific was asked to perform.

Given how cognitively advanced imitation is, you might be surprised to learn that unintentional imitation happens all the time! The chameleon effect, first described by Chartrand and Bargh (1999), involves unintentional imitation of body postures, facial expressions, and other behaviors of social partners. What's more, Chartrand and Bargh, as well as animal researchers, have found that when a social partner copies our behavior, we behave more positively toward them—that is, we like them more and prefer interacting with them more than noncopying social partners. In one study, two humans stood in front of an enclosure where a capuchin monkey had a ball. Each human had their own ball too but interacted with it differently. One played with their ball independently, while the other copied everything the monkey did. If the monkey chewed on their ball, the experimenter chewed her ball. If the monkey waved their ball in the air, the experimenter did too. The researchers found that in comparison to the noncopying human, the monkeys looked at the imitator more, spent more time in close proximity to the imitator, and wanted to play a token exchange game with the imitator more (Paukner, Suomi, Visalberghi, & Ferrari, 2009). As it turns out, imitation really is the highest form of flattery, and not just for humans!

As with many claims made by animal cognition researchers, critics will ask, "So what?" Why does it matter that a pond snail's behavior changes after observing another pond snail? Recent advances in neuroscience have shown that one of the main reasons why studying animals' social interactions is so important is that it teaches us something about how their brains change in response to social partners—arguably one of the most dynamic features of any animal's environment. Recently, neuroscientists discovered that when chimpanzees are taught to imitate humans via the "do as I do" method, a subset of the motor neurons in their brains actually change. Being trained to attend to and copy the behaviors of others caused the neurons to become more complexly interconnected with one another (Pope, Taglialatela, Skiba, & Hopkins, 2017). This is just one example that illustrates how sophisticated the processes associated with social interactions really are and how much we stand to learn from studying the influence of social partners on animals' behavior.

EMPATHY

To empathize is to take on the emotional state of another individual, to step into their shoes to feel what they feel, mainly as a result of having shared a similar prior experience. Because of empathy, we humans can modify how we interact with each other and behave in prosocial ways like helping, reassuring, reconciling, and bonding.

We often attribute the ability to empathize to our mammal and bird pets due to the depth of our interactions with them and the fact that most pets are social species to begin with.[2] However, empathy in animals is a hotly debated topic, in part because of empathy's inherent ties to emotion, which some see as uniquely human. There is also the difficulty in reaching consensus on a common definition for empathy. Some think we'll have to wait until neuroscience technology allows us to study empathy in the brain before we'll be able to define it (Knapska & Meyza, 2018).

For our purposes, we will use primatologists Frans de Waal and Stephanie Preston's (2017) definition of empathy, which is the process of being able to "understand others' states by activating [our] own personal, neural and mental representations of that state" (p. 498). They describe empathy as coming in two forms: affective empathy and cognitive empathy. Affective, or emotional, empathy involves witnessing the emotional state of another and matching it. For example, I might be so moved by seeing another in distress that I become emotional as well—connecting with the individual around their distress. It would not require action on my part, or even

---

2. Of all the feline species, only lions (*Panthera* spp.) and cheetahs (*Acinonyx jubatus*) are actually considered social. Domestic cats (*Felis catus*) vary in their preferred levels of sociality based on the environment they're in and how they're raised (see Shreve & Udell, 2015, for review).

necessarily a clear understanding of the situation. On the other hand, in cognitive empathy, I may see another in distress, think about a time when I was in that situation, appraise what I needed at that time in order to relieve my distress, and consider acting as a result of being informed by that cognitive appraisal. As is probably apparent, cognitive empathy requires cognitive and executive functioning abilities beyond simply matching the emotional state; however, both fall under the umbrella of empathy.

From an evolutionary stance, individuals who can empathize can both offer and be given support by group members, including cooperation, coordination of action, and the ability to respond to others in need (Clay, Palagi, & de Waal, 2018). All of these behaviors increase the likelihood of survival for any one individual in the group, making empathy a highly adaptive ability for many social species. On the other hand, for solitary species, conspecifics are often in direct competition, so empathy could negatively impact an individual's chance of survival.

More than any other topic in this chapter, empathy presents the biggest mystery to animal cognition researchers because complex internal motivation and intentionality must be inferred from rather simple outward behavior. Since the mid-20th century, physiological, behavioral, and neurological changes have been studied in both the laboratory and the field in order to draw conclusions about whether other animals can empathize. The prevailing view is that they can.

One of the first, if not the first, empathy studies focused on emotional contagion, the matching of a model's emotions (de Waal & Preston, 2017). Church (1959) trained a hungry observer rat to press a lever for food. However, every time the observer pressed the lever, it would cause another rat to be shocked. According to Church's logic, if rats are capable of affective empathy, we would expect to see behavioral indicators of distress in the observer, and that's just what he found. The hungry rat's lever pressing decreased, presumably because of some kind of emotional reaction to observing the other rat being shocked. In a similar study, monkeys chose to go over a week without eating rather than inflict pain on another monkey in order to access food (Wechkin, Masserman, & Terris, 1964). There are two possible interpretations here, each of which implies some kind of empathy. The first is that seeing the conspecific in pain caused a matching of the distressed animal's affective state and the observer became too emotionally distraught to eat. A more complex, cognitive explanation would be that the observer empathized with their conspecific who was in pain, "felt bad" for being the one causing the distress, and so stopped pressing the lever.

In addition to rats and monkeys, other nonhuman primates demonstrate emotional contagion, including mice (*Mus* spp.), chickens (*Gallus gallus*), and pigs (de Waal & Preston, 2017; Langford et al., 2006; Meyza, Ben-Ami Bartal, Monfils, Panksepp, & Knapska, 2017; Reimert, Bolhuis, Kemp, & Rodenburg, 2015). In each of these studies, observers interacted with a conspecific. What about interspecific emotional contagion? Canine cognition researchers have long wondered whether dogs feel what we feel. In one exploratory study, Custance and Mayer (2012) recorded dogs' behavioral responses to a talking, humming, or fake-crying human. Rather than approaching in an alert or playful way, dogs were more likely to approach the crying human in a submissive fashion and to nuzzle and lick them. The authors were careful to acknowledge that it did not necessarily imply the dogs empathized with the crying human, but rather that there was emotional state matching by the dogs of the human's presumed distress.

In the past, altruism (i.e., a donor offering some kind of help to another despite it being associated with some personal cost) was used interchangeably with empathy in psychological studies of both humans and nonhumans (e.g., Wolfle & Wolfle, 1939). Now, however, there appears to be agreement that altruism does not necessarily equate to empathy; rather, in order to demonstrate empathy, the donor must be motivated by "vicariously feeling the need of the recipient" (e.g., Silberberg et al., 2014, p. 610). As such, researchers more often use the term *helping behavior* to describe times when a donor acts so as to reduce the suffering or stress of another.

Helping behavior studies typically follow the same setup as emotional contagion studies, with the exception that the donor is given the opportunity to assist a social partner in some way to relieve its distress. This could come in the form of offering the observer the ability to press a lever to turn off an electrical shock, return a swimming rat to a dry platform, or release a partner trapped in a tight space. The first study of presumably empathetic helping behavior in animals was conducted by Wolfle and Wolfle in 1939 with monkeys, and since that time many other nonhuman primates have demonstrated the tendency to help others in need. Among nonprimates, studies referring to helping behavior in elephants, rats, dolphins, dogs, and humpback whales as based in empathy have been published (Bates et al., 2008; Ben-Ami Bartal, Decety, & Mason, 2011;

de Waal & Preston, 2017; Pitman et al., 2017; Quervel-Chaumette, Faerber, Faragó, Marshall-Pescini, & Range, 2016).

Another form of distress relief that may offer clues about animal empathy is consolation behavior, whereby following a distressing situation, an individual goes to help comfort a distressed social partner. As de Waal and Preston (2017) note, this form of helping behavior extends beyond basic reduction of pain or distress and is more "targeted" to the specific needs of the other, as when an observer takes on the perspective of another and offers their assistance. Consolation and reconciliation following conflict has been observed in a variety of species as well, including apes, monkeys, dogs, elephants, voles (*Microtus* spp.), mice, rooks (a species of corvid), rats, and parakeets (*Melopsittacus undulatus*) (Burkett et al., 2016; de Waal & Preston, 2017; Ikkatai, Watanabe, & Izawa, 2016; Webb, Romero, Franks, & de Waal, 2017).

The next time you are near a human with whom you share a strong emotional bond, casually yawn with your mouth open wide to give the full effect of the behavior. What you will likely find is that your social partner also yawns. Contagious yawning has been linked to areas of the brain that appear to be relevant to empathy (Platek et al., 2003). Truth be told, writing the word *yawn* so many times in this paragraph just made me (ECW) yawn. Not too sure what this says about my emotional bond with my computer.

Following the discovery of the link between contagious yawning and empathy in humans, some animal researchers decided to jump on the bandwagon and test it out. Chimpanzees, macaque monkeys, gelada baboons (*Theropithecus gelada*), dogs, and parakeets all yawn in response to seeing another yawn (de Waal & Preston, 2017; Yoon & Tennie, 2010). Interestingly, dogs will also yawn in response to seeing a human yawn (Joly-Mascheroni, Senju, & Shepherd, 2008). The relative ease with which contagious yawning can be studied makes it a favorite among those interested in assessing empathy in animals. The animal can watch a live model, a video clip, a human, a conspecific, an entirely different species, a familiar partner, or a stranger—the variations and combinations are limitless. Further, yawning is both involuntary and widespread throughout the animal kingdom. This has made comparative studies of evolutionarily similar and distant species possible, the results of which are invaluable to understanding the foundational building blocks of empathy. Still, critics point out the connection between empathy and yawning is correlational, meaning there may be more to the story than we think (Yoon & Tennie, 2010).

In addition to yawning, empathy is also correlated with two measurable neural responses: oxytocin release and mirror neuron activation. Sometimes discussed as a hormone, sometimes as a neurotransmitter, oxytocin is a chemical produced in the brain that is released during times of bonding, social support, trust, positive physical touch, and when inferring emotions and mental states of others. For decades, researchers have known that oxytocin is also connected to prosocial behavior in animals (Insel & Shapiro, 1992), making it a great candidate for corroborating animal cognition researchers' behavioral empathy findings. Oxytocin is typically studied in one of two ways: naturally or experimentally. Naturally, researchers can measure baseline levels of oxytocin in the blood, saliva, or urine, then subject an animal to an empathy-inducing task, then measure oxytocin levels after the experience to see if there was an increase. Another way to use oxytocin to study empathy is by manipulating oxytocin levels by introducing more oxytocin to the brain and seeing how it affects helping behavior or emotional contagion, for example.

First discovered in the macaque brain in the 1990s, mirror neurons are a type of neuron that are activated both when an animal engages in goal-directed actions such as grasping a cup and when the animal watches another individual engage in such actions. For example, if I were to watch you pick up a piece of food on a fork, put it in your mouth, and chew, the mirror neurons in my brain involved with those processes (picking up the food, putting it to your mouth, chewing) would activate as if I were the one actually engaging in those actions (Rizzolatti & Craighero, 2004). My neurons would be mirroring yours. But there's one caveat. If you picked up the food on your fork and flung it backward over your shoulder, no mirror neurons in my brain would be activated because that's not an expected motor action. I do not have pathways in my brain that normally activate for throwing food over my shoulder (Iacoboni & Dapretto, 2006). Along with empathy, mirror neurons are widely considered to be relevant to complex social processes such as theory of mind, culture, language, and imitation (Corballis, 2015; Iacoboni, 2009). Studying the mirror neuron system is inherently challenging due to the necessity to study neuron firing at an individual brain cell level; nonetheless, much can be learned about empathy in other species by looking at the brain.

So far, it seems that evidence coming from neuroscience supports what animal cognition researchers see in outward behavior, but good scientists always think in terms of alternative explanations. It's entirely possible that a rat or ape might help another not to decrease the other's distress but rather to reduce their own distress brought on by the social partner. In this way, the animal would be driven purely to self-soothe, rather than to be prosocial (de Waal & Preston, 2017). Likewise, what might look like a rat driven by empathy to release her cage partner from a tube could also be explained by the rat wanting her playmate back (Silberberg et al., 2014). I (ECW) once told a friend that her dog might lick her face when she's crying because he likes the flavor of her tears, not to cheer her up. That did not go over well. While these selfish, more parsimonious explanations for outwardly empathetic behavior might be hard to hear since we humans like to think animals feel with and for us, alternatives should not be ignored.

By designing appropriate control conditions to address these alternatives, researchers have concluded that many other species likely empathize. This begs the question of what it would mean if animals can truly "tap into" the emotional states of others. Empathy involves emotions, or at the very least complex affective states. While this might seem like a no-brainer for those who spend a lot of time with animals, the consequences of concluding animals have emotions would have far-reaching implications.

SOCIAL REFERENCING

As we saw in Chapter 4, Communication Between Animals, many animals can communicate using subtle social cues like eye gaze, pointing, and posture. One very special way humans and animals also use social cues is called *social referencing*, which involves learning from others' emotional responses. These responses, like a grimace after the first bite of a disgusting meal, act as signals that communicate information to social partners. For example, when a young child falls down and immediately looks up at her caregiver to see their reaction, the child gains information about how to respond to the situation. Social referencing interactions like this allow one individual to look to a social partner and appraise their reaction toward a shared object or event. This gives them information about how to appropriately respond emotionally and behaviorally (Walle, Reschke, & Knothe, 2017). If the caregiver looks terrified, the child will likely cry harder than if the caregiver downplays the child's fall (e.g., "You're okay. Get up."). This seeking out of information from others' emotional cues makes social referencing distinctly different from other nonverbal communication cues (Walle et al., 2017).

In humans, the ability to socially reference others emerges early in life, is refined over time via experience, and is crucial to productive social interactions and survival (Klinnert, Campos, Sorce, Emde, & Svejda, 2013; Walden & Ogan, 1988). This seems to be the case for other primates as well. When it comes to avoiding potentially poisonous foods, for example, infant chimpanzees will look to their mothers for approval before tasting something they've never eaten before, but they do not do this with familiar foods (Ueno & Matsuzawa, 2005). Fear can also be learned via social referencing. In a famous comparison between laboratory- and wild-born infant macaques, both groups developed an intense fear of real and even toy snakes after seeing their wild-born parents respond fearfully to them (Mineka, Davidson, Cook, & Keir, 1984). Importantly, none of the infants were afraid of snakes before observing their parents, indicating that their parents' reaction offered new information about how to interpret the world. As these examples illustrate, an animal who is willing to look to others and change their behavior in response to what they see gets to learn a valuable lesson without having a painful (or potentially deadly) firsthand experience.

While there are some conspecific social referencing studies out there, for practical reasons it's easier to test how animals' react to humans' emotional responses. The results of these studies have shown that social referencing is so evolutionarily adaptive that some species readily look to humans for more information (Heyes & Galef, 1996). For example, young chimpanzees and cats socially reference humans when they're presented with a novel object before interacting with it (Merola, Lazzaroni, Mashall-Pescini, & Prato-Previde, 2015; Russell, Bard, & Adamson, 1997). In Merola et al.'s study, some of the cats even alternated between looking to the human and back at the object three or more times, as if to repeatedly update their understanding of the human's emotional reaction.

In addition to presenting novel objects, one commonly used social referencing design is called the object-choice task. If two identical objects are presented, it's in the subject's best interest to pay attention to how another individual responds to them before deciding which one to pick. Morimoto and Fujita (2012) tested this by training a demonstrator capuchin monkey to produce either a positive or negative reaction to two identical boxes. As predicted, observer monkeys were more likely to investigate boxes that demonstrators responded positively toward. This indicated to the authors that the observers, unsure of which box to choose, looked to their partner for more information, which then informed their behavior.

Among all animals, dogs are the best at responding to human-delivered cues, even outperforming chimpanzees on an object-choice task that used pointing (Kirchhofer, Zimmermann, Kaminski, & Tomasello, 2012). When it comes to social referencing specifically, dogs can respond to their owners' facial expressions (Merola, Prato-Previde, & Marshall-Pescini, 2012) and are also sensitive to images of strangers' happy or angry faces and the tone of their words (Albuquerque et al., 2016). Testing dogs' social referencing ability at an even more nuanced level, my (ECW) laboratory has shown that dogs' object-choice task behavior is affected by an interaction between pointing gestures and nonverbal social referencing vocalizations (Colbert-White, Tullis, Andresen, Parker, & Patterson, 2018). Dogs saw us put a piece of cheese inside one of the two containers and mix them up. Without any information to go off (we controlled for the cheese smell by secretly having cheese inside both), the dogs' best bet was to see how the human responded when she looked inside each container. When she looked inside one of the containers and made a happy sound, the subjects were much more likely to choose that container, as if they had inferred from her reaction that the cheese must be inside it.

If an animal socially references a conspecific or a human, it suggests they view them as a social partner worth learning from, similar to our previous discussion of social and observational learning. Likewise, if an animal learns by interpreting another's facial expressions, it implies the animal may recognize the internal state of the other, trust that information, and respond in an informed way from what they learned from the social partner. There are many assumptions that must be made in order to conclude that social referencing has occurred, including the idea that animals have emotions, and that animals can make inferences about others' internal states, which starts to get into the realm of theory of mind. Critics argue that rather than some innate emotion-sharing experience, an animal that backs away from a novel object that another animal reacted toward could simply be responding via basic conditioning. For example, a cat may have learned from previous experiences that when the owner makes a particular face, something bad is going to happen, so it's best to avoid that object. This interpretation does not require the cat to have a built-in special human emotion decoder. Rather, animal social referencing in the wild and in the laboratory could be a reflection of very nuanced associative learning. Nonetheless, there does appear to be a pattern—typically social mammals excel at social referencing compared to solitary animals—which we would not necessarily expect if social referencing performance could be explained away as basic conditioning.

ANIMAL CULTURE

It's easy to take for granted that many of our thoughts, ideas, and behaviors are not uniquely ours. Rather, they're representations of norms from our families, communities, and geographic locations. As we begin this final topic of social cognition, we invite you to take a few minutes to respond to the following brief exercise. The questions are designed to unpack some of your ideas and expectations about culture—an abstractly defined yet frequently used word.

1. How do you define culture?
2. With which culture do you identify the most?
3. What are some indicators to you that you identify the most with this culture?

If you noticed you did not immediately have an answer to some or all of the questions, that's okay! It just means that you are thinking about culture more deeply than you had previously. Now, keeping your responses in mind, let's delve deeper into the debate of whether animals have culture.

First and foremost, what does it mean that humans have culture? Looking to sociology, anthropology, and psychology, there are some differences in definitions, but by and large, the disciplines agree on some fundamental similarities. For our purposes, we have selected a definition for culture that encompasses many of these similarities. In his book on cultural psychology, Heine (2011) defines culture as "any kind of information that is acquired from other members of one's species through social learning that is capable of affecting an individual's behavior" (p. 3). This is essentially saying that culture is a set of behaviors we copy from those we spend a lot of time with. Looking at the definition you wrote earlier, you may see similarities to Heine's. You may also notice that the indicators you listed for your own culture—whether it be food, dress, customs, or skill sets—are reflected in Heine's definition as examples of the kind of information that is passed down from generation to generation.

At the risk of sounding like a broken record, animal culture is highly controversial among scholars. Some anthropologists and psychologists strongly believe that culture is a unique hallmark of humanity. These folks claim culture is one of the main steps our ape-like ancestor evolved on the road to humanness, along with tool use, bipedalism, reduced canine tooth size, hunting, language, and domestication (e.g., Larsen, 2017). Others have a more inclusive view of culture and are open to the possibility that culture is one of the steps to humanness that animals also share (e.g., McGrew, 2004). The general consensus following more than half a century of research is that there is evidence for animal culture, though it seems to appear in less complex forms than it does in humans (Claidière & Whiten, 2012). Most anthropologists and psychologists have accepted that chimpanzees have culture. Beyond this, culturally transmitted behaviors have also been observed in monkeys, other mammals, birds, fish, and even insects. Thinking from an adaptive stance, it would make sense for culture to have evolved multiple times in the animal kingdom. Members from highly social, group-living species (particularly those with long periods of parental care and long life spans) have more time to spend observing one another and sharing what they've learned with others, which in turn increases survival and group cohesion.

Studying animal culture takes time, a knowledge of individual members in the focal group, as well as a keen eye for observing behavior. At some point, a new method of foraging or a leisure activity like snowball rolling (a pastime of some macaques!) is either trained by a human experimenter or spontaneously learned by a select number of individuals, and then the extent to which the behavior "catches on" and persists among members in the larger group is recorded. Conclusions drawn from studying these culturally transmitted behaviors must be evaluated with caution, as some animal cognition researchers argue these behaviors do not necessarily imply full-blown culture (e.g., Galef, 1992). Nonetheless, by definition, there appears to be some connection between culturally transmitted behaviors and culture.

Animal culture can be studied in multiple locations. Some choose to look for culturally transmitted behaviors in captive animals by teaching subsets of group members two different ways to solve a food puzzle and then tracking how the methods are passed down vertically from parent to offspring and horizontally from peer to peer (e.g., Crast, Hardy, & Fragaszy, 2010). Others try their luck looking for cultural transmission of behavior in the wild. Similar to how a cultural anthropologist might study human cultures, field animal culture researchers like the famed primatologist Jane Goodall spend hundreds, even thousands, of hours immersed with a group of animals looking for evidence of one animal's novel behavior spreading throughout the group. Lawick-Goodall's (1968) longitudinal (i.e., lasting over a period of time) research with wild chimpanzees documented a number of innovative foraging techniques being passed down, like dipping twigs into termite mounds. Another field study of wild chimpanzees noted the idiosyncratic culturally transmitted behavior of using clumps of moss to soak up water to drink (Lamon, Neumann, Gruber, & Zuberbühler, 2017).

Just like human cultures differ on a number of dimensions, if animals have culture, we should be able to find distinct behavioral differences between groups of the same species that cannot be explained by biology (Claidière & Whiten, 2012). And we do. Humpback whales' songs vary geographically by group, and to the surprise of researchers, over the course of only 2 years, one group adopted the song of visitors from a different geographical region (Noad, Cato, Bryden, Jenner, & Jenner, 2000). Hunting and foraging behaviors, novel play behaviors among young dolphin peers (Kuczaj, Makecha, Trone, Paulos, & Ramos, 2006), cooperative behavior with humans, as well as other examples of culturally transmitted behaviors and dialects are also widespread across individual groups of dolphins, orcas, belugas (*Delphinapterus leucas*), and other cetacean species (we direct the reader to Rendell & Whitehead's extensive review from 2001, which also includes many more definitions of culture).

While the field is a great place for those patient few who are willing to wait to find spontaneous examples of animal traditions (i.e., a specific behavioral pattern that is socially learned, shared by others in the group, and present over a long span of time; Fragaszy & Perry, 2003), many others opt to artificially induce traditions and see how they are transmitted via social learning. For example, de Waal, Borgeaud, and Whiten (2013) conducted a clever study in which vervets monkeys' (*Chlorocebus aethiops*) food preferences were experimentally manipulated. One group was given the choice between blue-dyed and pink-dyed corn, where the blue was made to taste very bitter. The group learned quickly to only consume the pink corn. Naturally, the next generation of infants selectively ate pink corn like their mothers. But what's really interesting is that when bachelor males from a second group where everyone ate blue corn (pink was made to taste bitter) traveled to the pink corn group, they conformed to the group and began eating pink corn like everyone else. We might consider this the vervet monkey equivalent of moving to a new country and abandoning your previous culture's way of life because of the strong social influence and immersion in the new culture. Like this and Crast et al.'s (2010) dual-puzzle-solving method, these kind of lab-based tradition studies offer controlled, rigorous evidence that group-level differences within a species can be explained by experience rather than biology.

Most examples of animal culture come from observational studies in which scientists record how a behavior is passed down by watching individual animals. Thanks to biomolecular methods, in many cases, behavioral observations can now be corroborated using genetic testing. For example, dolphins in Shark Bay, Western Australia, use sponges to stir up sediment and small prey on the ocean floor. In this population, mothers appear to actively teach sponging behavior to their calves, which can be observed directly, and now we can confirm it via tracing family lines with matrilineal DNA (Kopps et al., 2014; Krützen et al., 2005). Genetic testing can help researchers who arrive at a new field site without any of the valuable history of a population determine who most likely reared whom, who is not related to the main family group but has likely been a part of the group for awhile since their offspring share DNA with the main family group, and who is completely unrelated and just showed up one day with a new way of doing things. Would a youngster be more curious about and therefore more likely to learn from a stranger, from his peers, or from his parents? These are all questions that can now be answered with the help of genetic testing.

Unlike nonhuman primates and cetaceans, less attention has been paid to studying culture in birds, fish, and insects. Songbirds, parrots, and corvids are among the most commonly studied birds, likely due to their complex sociality and sophisticated vocal repertoires. Research has shown that ravens and parrots' repertoires are culturally distinct "dialects" with some shared overlap (i.e., cultural transmission) of vocalizations from neighboring conspecific groups

(Aplin, 2019; Enggist-Dueblin & Pfister, 2002). Further, similar to what has been observed with monkeys (e.g., Crast et al., 2010), songbirds can also pass down a tradition for solving a puzzle box using whichever of two solving methods they observed from a model (Aplin et al., 2015). In the case of Aplin et al.'s birds, they adapted in response to their social group, and they actually preferentially adopted the information passed down via cultural transmission rather than relying on what they had learned themselves.

While the preceding examples largely came from communication and problem-solving literature, tool use, foraging methods, and predator recognition can also be passed down via cultural transmission. Some social fish species learn from group members which animals are dangerous. For example, naive minnows (*Phoxinus phoxinus*) learn from experienced minnows to avoid pike (*Esox lucius*), a predatory fish, and will seek shelter though they've never personally had a negative experience with pike before (Mathis, Chivers, & Smith, 1996). As we discussed much earlier in the chapter, being able to learn from others without trial and error is highly adaptive and can be the difference between life and death. But what about when culture leads a group down a path of being less adaptive, such as when a particular tradition or behavior that has been passed down through generations is either no longer helpful or actually harmful? Laland and Williams (1998) observed this very thing when they trained a handful of guppies (*Poecilia reticulate*) to either take an easy route to get to a food source or a roundabout, energy-wasting path. Once trained, the "teacher" fish were released into the rest of the group, and the researchers observed who learned what from whom. Some did adopt the inefficient route and continued to use it after the teachers were removed from the group. Those inefficient fish also clung to their inefficient method and took longer to learn the easier route. Like many humans, it seems fish are not immune to the problems associated with following the herd!

Considering the controversy of animal culture, we could not end this section without sharing an exciting study on insects. Eusocial species like bees, ants, and wasps are highly social. You may have heard the term super-organism to describe how cooperatively they work together to create a functioning nest. Recently, bumblebees demonstrated the ability to learn via observation to solve the string-pulling task, where an out-of-reach reward attached to a string can only be obtained by pulling the string (Alem et al., 2016). Alem and colleagues also showed that this behavior was passed down horizontally in the group by multiple "generations" of newly added naive bees, indicating the behavior was both socially learned and passed down—which is our working definition of culture.

Despite differences in definitions, one thing appears certain: Culture requires a collection of socially acquired information that persists through time. In humans, this involves intentional teaching by models and adoption of customs, traditions, dress, skills, and so on by new members. This assimilation allows cultures to thrive and, in some cases, remain virtually unchanged. As discussed at the beginning of this chapter, there is no clear consensus on whether animals teach each other, which casts some doubt on the similarity of their behavior to human culture. Nevertheless, the animal kingdom is full of examples of what could easily be called the building blocks for culture, which makes their behavior equally intriguing.

ANIMAL SPOTLIGHT: IMO

Of all the animal spotlights, Imo may be the least well known, even to the most dedicated animal cognition fan. This social cognition chapter was organized by increasing complexity of sociocognitive abilities. Moving from learning by observing, to matching emotional states, to learning from others' emotional states, and finishing with cultural transmission of behavior across future generations. For this spotlight, we highlight Imo, a female Japanese macaque (*Macaca fuscata*) who was the original focal animal in the longest-running project on animal culture.

The small island of Koshima (Kōjima, 幸島) is located off the coast of the Miyazaki Prefecture in southern Japan. In the late 1940s, the Japanese Primate Research Institute (PRI) established the island as a field site. Researchers Kinji Imanishi, Shunzo Kawamura, and Junichiro Itani framed the initial project as a longitudinal study of social behavior of the wild macaques living on the island (Hirata, Watanabe, & Kawai, 2001). The group chose to give each animal an individual name so they could better follow the dynamics of the group's social interactions, which was one of the main goals of their project. With the exception of naming the animals and periodically leaving food out to attract them to their research area, the monkeys were minimally disturbed by humans. This tradition has remained the same for more than 60 years.

In 1953, a 1.5-year-old macaque named Imo (Japanese for *potato* or *yam*) surprised the PRI researchers when she grabbed a chunk of sweet potato off the beach and began rinsing the sand off in a freshwater brook before eating it (Kawamura, 1954, in Japanese). Until then, all of the other monkeys would brush as much sand off as possible using their hands. Imo's faster and more effective method spread throughout the group—first to her playmate Semushi after 1 month, then to her mother Eba and another playmate Uni (Galef, 1992).

The spread of Imo's sweet potato washing supported Imanishi's (1952, in Japanese) early hypothesis that nonhuman primates possessed the behavioral building blocks of culture. Kawamura (1954) presented the first account of Imo's behavior in Japanese, and Kawai (1965) retold the story for a wider English-speaking audience a decade later, catapulting Imo into primate stardom. Because all of the animals could be identified, the PRI's publications could note interesting demographic findings. For example, adults over 8 years old were less likely to imitate Imo's sweet potato washing behavior compared to youngsters, who readily adopted it from Imo, her mother, and her siblings. There were also sex differences, with adolescent and adult females being much more likely to adopt Imo's methods than males. The authors explained this difference based on the natural social organization of Japanese macaques. Males are distanced from the entire social group once they hit adolescence, so they'd be less likely to pick up the behavior. These demographic findings further illustrated the fascinating spread of the behavior. Specifically, it was not just about how much time you spent with Imo; other demographic factors contributed to whether or not you picked up the sweet potato washing behavior.

But Imo was not done showing off her ingenuity! While sweet potato washing was impressive, Imo took it to another level by figuring out how to season them as well. Eventually, she discovered that sweet potatoes rinsed in the ocean were a more flavorful snack than those rinsed in the freshwater. So, after each bite, she dipped the potato back into the ocean. At some point, Kawai (1965) predicted between 1954 and 1957, many of the monkeys switched over from using the river to using the salty ocean as well, and by 1961, almost everyone was using the ocean. While Kawai acknowledges the move could have been because of the abundance of salt water compared to fresh water, the passing down of flavoring the potatoes was considered a more likely reason.

Following success with sweet potato washing and seasoning, trendsetter Imo showcased her unique innovation once again at 4 years old when she spearheaded wheat washing. The PRI scattered wheat grains onto the beach for the monkeys, who normally picked the grains out of the sand one by one. Imo was not interested in this tedious process, so she instead dropped handfuls of wheat–sand mix into the water, let the sand sink, then skimmed the floating wheat grains off the top. This behavior did not initially spread across and down lineages as quickly as the sweet potato washing and seasoning did, but after the PRI researchers really stamped the wheat into the sand, other monkeys eventually took up Imo's washing method.

While some might say the PRI researchers were unscientific to interfere to speed up Imo's socially transmitted wheat washing behavior, from a scientific standpoint, they did the right thing as researchers by reporting what they had done. Without their careful methodological records, someone might have mistakenly believed the behavior spread on its own. For groups of Japanese macaques living in a colder area of the country, decades of assumptions about one of their well-known traditions revealed a disappointing inaccuracy. These monkeys bathe in the natural hot springs, and people thought the tradition began with one creative individual who tried it out first and others following suit, just like in Imo's case. Digging deeper, we now know that the earliest groups of Japanese macaques were actually rewarded with food by humans for going into the hot springs. This would be like someone training the chimpanzees to fish for termites with sticks days before Jane Goodall showed up with her research team. The point of the story is this: Even "wild" animals may not be immune to human intervention. Since they do not have any detailed historical records of the animals, field researchers who stumble upon socially transmitted behaviors in the wild must be very careful in their investigations (Tomasello, 1996).

Kawai's (1965) summary of Imo's behavior has been cited almost 900 times by psychologists, anthropologists, and others. What is particularly fascinating about both sweet potato washing and wheat washing is the fact that Imo's innovations involved using water as a tool, and tool use was long considered to be uniquely human. Thus, when Imo's story was first published, some psychologists and anthropologists were particularly ruffled by the claims of both tool use *and* culture. Adding another layer of complexity, these behaviors were acquired without language or active teaching (Kawai, 1965).

Some point out that Imo is not the only Japanese macaque to spontaneously wash potatoes and many other monkey species have learned to wash their food. Further, because the monkeys were given food by humans, there is always the possibility that a human inadvertently reinforced a behavior beyond the recorded monkey-to-monkey transmission (Galef, 1992). Nonetheless, the story of Imo and the monkeys of Koshima island, as well as the persistence of their food culture, has greatly informed the area of animal culture.

Though Imo and her original community of Japanese macaques are no longer alive today, the monkeys on the island still clean and season their food using the methods Imo first developed. To put her innovations in perspective, imagine a human child spontaneously inventing a more efficient way to process food and that method being adopted by the whole town for generations to come! Thinking of Imo's behavior that way, you can hopefully understand why we chose to highlight her. Not only did she stand out within her group, but her legacy lives on decades later, providing some of the most compelling evidence for animal culture.

HUMAN APPLICATION: TAGTEACH AND OPERANT CONDITIONING

Animal trainers say that it's possible to get any animal to do anything—you just have to have the right reinforcement. Karen Pryor (1999), in her book *Don't Shoot the Dog*, mentions a professor who once claimed he had taught a scallop to clap for food. If you've got a goldfish and some spare time, check out the R2 Fish Training School—you will not be disappointed.

Using reinforcement or punishment to change an organism's behavior is called *operant conditioning*. The terminology can be a little complicated—in operant conditioning, anything that makes a behavior more likely in the future is a reinforcement, and anything that makes it less likely is a punishment. That's pretty straightforward. A cookie for your dog after he sits is a reinforcement; a tap on the nose after he jumps on a stranger is a punishment. Reinforcement and punishment are also directional, if you will. In operant conditioning, anything that is added to the situation is positive, while anything removed is negative. So giving your dog that cookie is positive reinforcement. You added something (the cookie) that makes the behavior (sitting) more likely in the future. But that tap on the nose? Positive punishment. You added something to the situation (the tap) in order to make the behavior (jumping) less likely in the future. Letting your dog out of her crate because she had been quiet? Negative (removal of the space restriction) reinforcement (to increase the chance of quiet behavior in the future). Taking away her bone for growling at another dog? Negative (removal of the bone) punishment (to decrease the chance she'll growl in the future).

While there is a time and place for everything, studies have shown repeatedly that positive reinforcement is by far the best method for long-term behavioral change while maintaining a positive relationship between teacher and learner (Premack, 1959; Sigler & Aamidor, 2005). In addition, teachers are encouraged to avoid building negative associations—tapping your dog often might lead him to cringe whenever he sees you, even if he's not in trouble. Highlighting and rewarding what you want to see your learner do also communicates what the correct behavior is. Remember when the elementary school teachers yelled "Walk in the halls" instead of "Don't run in the halls"? There was a reason for that.

In fact, much of what we do every day can be broken down to a series of social interactions that are grounded in operant conditioning. The mother ignoring her child's tantrum in the supermarket? Negative punishment (removal of attention to decrease the likelihood of tantruming). When visiting a foreign country, a cultural faux pas may earn you a scowl or stern look to decrease the chance you'll do the same thing next time (positive punishment), but a gift for your hosts will increase the chance they'll invite you back to their home (positive reinforcement). By reinforcing and punishing behaviors, social norms can be learned by all community members.

Taking inspiration from B. F. Skinner and his pigeons, some educators and trainers have begun to harness the principles of operant conditioning to help people learn in applied settings like schools and athletics. By using a tool called a "bridge," it is possible to reinforce an exact behavior right as it's being done. A bridge—often a clicker or a whistle—means "That's it right there; now come get your reinforcement." As the name implies, the bridge bridges the gap between behavior and reinforcement delivery.

In athletics, coaches can use this to identify specific movements or motions (Fogel, Weil, & Burris, 2010; Quinn, Miltenberger, & Fogel, 2015). For example, say a gymnast is working on keeping her toes pointing in a particular spot in the routine—it's great for her coach to tell her "you got it" after she finishes, or even yell the phrase during the routine. But a sharp click or whistle at the exact moment she points her toes correctly allows her to identify the position and feel of her body right then. Or, perhaps, a golfer is working on keeping her arm bent a specific amount when swinging the club—the bridge allows the athlete and the coach to narrowly focus on the arm angle specifically before returning to the mechanics of the swing overall.

A movement called TAGteach, or Teaching with Acoustical Guidance (TAGteach International, 2004), has had overwhelmingly positive results using a bridge to teach children and adults everything from handwriting to meat processing in assembly lines. By setting "tag points" to break down the goal behavior into smaller behaviors, learners find it easier to manage. For example, writing the letter M requires making four lines. Or it involves making one line four times—each time followed by a click from the teacher to note you're on the right track. Mastering the necessary cuts and order of processing meat at a fish cannery, for example, can be overwhelming to learn all at once. It's much more manageable if the employee first focuses on mastering one major step (click!), then another (click!), then the next (click!), and so on until the whole behavior can be chained together as one sequence.

Because the principles TAGteach relies on are very basic and easily conveyed, it has become very useful working with people with autism spectrum disorder and has even been used to help teach developmentally disabled children important social and behavioral skills (LaMarca, Gevirtz, Lincoln, & Pineda, 2018). In one case, audio feedback with clicks helped correct toe walking in a child with autism, a behavior that can be very difficult to change (Persicke, Jackson, & Adams, 2014). Moreover, in a case that is highly relevant for anyone who has undergone surgery, TAGteach has been used by medical school professors to increase the precision by which surgeons learn specialized techniques. In one experiment, medical students who learned two new surgical techniques via TAGteach were more precise with their new skills than students who were taught via traditional teaching methods (Levy, Pryor, & McKeon, 2016). It sounds wild—applying the same methods dog trainers use to future surgeons—and in fact, the surgical techniques study was profiled in *The Annals of Improbable Research* (motto: Research that makes people LAUGH and then THINK). But hey, if TAGteach makes better surgeons and helps those with autism spectrum disorder, we're all for it!

YOUR TURN!

While we humans learn from one another all the time, we may not notice it. Likewise, even if you've seen the consequences of your dog learning how to open the pantry door to get to her kibble, you might not have given much attention to how she learned it. Did she learn it from you, or did she learn it on her own via trial and error? In this ministudy, we'll look more closely at the extent to which animals in human care learn directly from our teaching.

You'll need a stopwatch and a food-motivated animal you are very familiar with and with whom you can directly interact. You'll also need to be able to create an ethogram, which we explained in detail in Chapter 2, Theoretical and Methodological Approaches to Animal Cognition. Finally, you will need a puzzle that is solvable with minimal steps and whose solution can be clearly demonstrated to the animal by you. Be creative with whatever materials you have around the house. Animal cognition research does not have to be fancy or glamorous. A basic shoebox with a lid would be perfect for a medium-sized dog, for example.

We'll do two trials. The first trial involves collecting some baseline data on how your subject generally interacts with the puzzle. Take one of your animal's favorite treat, and bait your puzzle out of view. When you're ready, set the timer, and construct a 1-minute baseline ethogram to record data on what your animal does when you set the puzzle down in front of them. If they solve the puzzle and get the treat, that's great, but it means back to the drawing board for you! You want something that the animal presumably cannot solve without watching you solve it first.

As you do your baseline observation, there are a few key behaviors you might look for. One of those is the frequency and kind of vocalizations they emit. You might also be mindful of the animal's general activity level at the beginning of the minute compared to the end of it. For instance, if your subject goes longer without being able to solve the puzzle, they may become disinterested or move more slowly. Third, note specifically how your subject interacts with the puzzle, with attention paid to where on the puzzle they interact with it. Finally, does your subject use social cues like looking at the puzzle, then to you, and then back to the puzzle? Do they behave as if they're trying to solicit help from you? Be as neutral and unobtrusive as possible as you observe. This could be especially challenging if your animal tries to interact with you. Be strong and resist sad kitty eyes for the sake of science.

After the minute is up, move on to the second trial. Start by getting your animal's attention, and then solve the puzzle while they observe you. You can repeat the solution a few times if you'd like. Set the timer, place the secured puzzle back down on the floor, and begin part two of your ethogram—carefully tallying original behaviors and making note of new behaviors and their frequencies. Here is where the different behaviors in the previous paragraph may change. There may be some evidence of local enhancement, whereby the places on the puzzle you focused on with your hands receive more sniffing or greater attention, for example. Your subject may look to you less often, vocalize differently, or be more engaged now that they have seen the puzzle solved.

After the minute is up, if they still have not solved it, you could try demonstrating the solution again and observing for another minute. If they still have not solved it, give them a treat for being a great participant, and take a break. Ask yourself about the appropriateness of the task to the species—which might require some outside reading. Maybe the required lifting up of a shoebox flap is not a behavior that comes naturally to that species for some anatomical reason, which is why you have not observed that behavior at all. If you learn something that might be helpful, go back to the drawing board, develop a more species-appropriate puzzle, then try again. Maybe in your reading you find that the species you're testing does not seem to view humans as social partners they can learn from, which might explain your results. Whether they solve the puzzle or not, this kind of interaction is great cognitive stimulation for the animal and a fun application of what you've learned about social cognition.

REFERENCES

Albuquerque, N., Guo, K., Wilkinson, A., Savalli, C., Otta, E., & Mills, D. (2016). Dogs recognize dog and human emotions. *Biology Letters, 12*, 20150883. doi:10.1098/rsbl.2015.0883

Alem, S., Perry, C. J., Zhu, X., Loukola, O. J., Ingraham, T., Søvik, E., & Chittka, L. (2016). Associative mechanisms allow for social learning and cultural transmission of string pulling in an insect. *PLoS Biology, 14*, e1002564. doi:10.1371/journal.pbio.1002564

Aplin, L. M. (2019). Culture and cultural evolution in birds: A review of the evidence. *Animal Behaviour, 147*, 179–187. doi:10.1016/j.anbehav.2018.05.001

Aplin, L. M., Farine, D. R., Morand-Ferron, J., Cockburn, A., Thornton, A., & Sheldon, B. C. (2015). Experimentally induced innovations lead to persistent culture via conformity in wild birds. *Nature, 518*, 538–541. doi:10.1038/nature13998

Avarguès-Weber, A., & Chittka, L. (2014). Local enhancement or stimulus enhancement? Bumblebee social learning results in a specific pattern of flower preference. *Animal Behaviour, 97*, 185–191. doi:10.1016/j.anbehav.2014.09.020

Avarguès-Weber, A., Lachlan, R., & Chittka, L. (2018). Bumblebee social learning can lead to suboptimal foraging choices. *Animal Behaviour, 135*, 209–214. doi:10.1016/j.anbehav.2017.11.022

Bandura, A. (1976). *Social learning theory*. Englewood Cliffs, NJ: Prentice-Hall.

Bates, L. A., Lee, P. C., Njiraini, N., Poole, J. H., Sayialel, K., Sayialel, S., ... Byrne, R. W. (2008). Do elephants show empathy? *Journal of Consciousness Studies, 15*, 204–225.

Bem, T., Jura, B., Bontempi, B., & Meyrand, P. (2018). Observational learning of a spatial discrimination task by rats: Learning from the mistakes of others? *Animal Behaviour, 135*, 85–96. doi:10.1016/j.anbehav.2017.10.018

Ben-Ami Bartal, I., Decety, J., & Mason, P. (2011). Empathy and pro-social behavior in rats. *Science, 334*, 1427–1430. doi:10.1126/science.1210789

Berg, K. S., Delgado, S., Cortopassi, K. A., Beissinger, S. R., & Bradbury, J. W. (2012). Vertical transmission of learned signatures in a wild parrot. *Proceedings of the Royal Society B, 279*, 585–591. doi:10.1098/rspb.2011.0932

Boughman, J. W., & Wilkinson, G. S. (1998). Greater spear-nosed bats discriminate group mates by vocalizations. *Animal Behaviour, 55*, 1717–1732. doi:10.1006/anbe.1997.0721

Brown, C., & Laland, K. N. (2003). Social learning in fishes: A review. *Fish and Fisheries, 4*, 280–288. doi:10.1046/j.1467-2979.2003.00122.x

Burkett, J. P., Andari, E., Johnson, Z. V., Curry, D. C., de Waal, F. B. M., & Young, L. J. (2016). Oxytocin-dependent consolation behavior in rodents. *Science, 351*, 375–378. doi:10.1126/science.aac4785

Byrne, R. W., & Whiten, A. (1988). *Machiavellian intelligence: Social expertise and the evolution of intellect in monkeys, apes and humans*. Oxford, UK: Oxford University Press.

Caldwell, M. C., & Caldwell, D. K. (1965). Individualized whistle contours in bottle-nosed Dolphins (*Tursiops truncatus*). *Nature, 207*, 434–435. doi:10.1038/207434a0

Chapman, T. L., Holcomb, M. P., Spivey, K. L., Sehr, E. K., & Gall, B. G. (2015). A test of local enhancement in amphibians. *Ethology, 121*, 308–314. doi:10.1111/eth.12337

Chartrand, T. L., & Bargh, J. A. (1999). The chameleon effect: The perception-behavior link and social interaction. *Journal of Personality and Social Psychology, 76*, 893–910. doi:10.1037/0022-3514.76.6.893

Church, R. M. (1959). Emotional reactions of rats to the pain of others. *Journal of Comparative Physiological Psychology, 52*, 132–134. doi:10.1037/h0043531

Claidière, N., & Whiten, A. (2012). Integrating the study of conformity and culture in humans and nonhumans animals. *Psychological Bulletin, 138*, 126–145. doi:10.1037/a0025868

Clay, Z., Palagi, E., & de Waal, F. M. B. (2018). Ethological approaches to empathy in primates. In E. Knapska & K. Z. Meyza (Eds.), *Neuronal correlates of empathy: From rodent to human* (pp. 53–66). London, UK: Elsevier.

Clutton-Brock, T. H., Russell, A. F., & Sharpe, L. L. (2004). Behavioural tactics of breeders in cooperative meerkats. *Animal Behaviour, 68*, 1029–1040. doi:10.1016/j.anbehav.2003.10.024

Colbert-White, E. N., Tullis, A., Andresen, D. R., Parker, K. M., & Patterson, K. E. (2018). Can dogs use vocal intonation as a social referencing cue in an object choice task? *Animal Cognition, 21*, 253–265. doi:10.1007/s10071-018-1163-5

Corballis, M. (2015). Theory of mirror neurons. In J. Wright (Ed.), *International encyclopedia of the social & behavioral sciences* (pp. 582–588). Oxford, UK: Elsevier.

Cracknell, N. R., Mills, D. S., & Kaulfuss, P. (2008). Can stimulus enhancement explain the apparent success of the model-rival technique in the domestic dog (*Canis familiaris*)? *Applied Animal Behaviour Science, 114*, 461–472. doi:10.1016/j.applanim.2008.04.004

Crast, J., Hardy, J. M., & Fragaszy, D. (2010). Inducing traditions in captive capuchin monkeys (*Cebus apella*). *Animal Behaviour, 80*, 955–964.

Custance, D., & Mayer, J. (2012). Empathetic-like responding by domestic dogs (*Canis familiaris*) to distress in humans: An exploratory study. *Animal Cognition, 15*, 851–859. doi:10.1007/s10071-012-0510-1

Davis, K. M., & Burghardt, G. M. (2011). Turtles (*Pseudemys nelsoni*) learn about visual cues indicating food from experienced turtles. *Journal of Comparative Psychology, 125*(4), 404–410. doi:10.1037/a0024784

de Waal, F. B. M., & Preston, S. D. (2017). Mammalian empathy: Behavioural manifestations and neural basis. *Nature Reviews Neuroscience, 18*, 498–509. doi:10.1038/nrn.2017.72

Duffy, J. E. (2003). The ecology and evolution of eusociality in sponge-dwelling shrimp. In T. Kikuchi, S. Higashi, & N. Azuma (Eds.), *Genes, behavior, and evolution in social insects* (pp. 217–254). Sapporo, JPN: Hokkaido University Press.

Enggist-Dueblin, P., & Pfister, U. (2002). Cultural transmission of vocalizations in ravens, *Corvus corax*. *Animal Behaviour, 64*, 831–841. doi:10.1006/anbe.2002.2016

Fiorito, G., Rosciano, V., Scotto, P., & Valsecchi, P. (1994). Assessing observational learning in *Octopus vulgaris*: The jar problem solution. *Social Behaviour, 61*, 50. doi:10.1080/11250009409355988

Fiorito, G., & Scotto, P. (1992). Observational learning in *Octopus vulgaris*. *Science, 256,* 545–547. doi:10.1126/science.256.5056.545

Fogel, V. A., Weil, T. M., & Burris, H. (2010). Evaluating the efficacy of TAGteach as a training strategy for teaching a golf swing. *Journal of Behavioral Health and Medicine, 1*(1), 25–41. doi:10.1037/h0100539

Fragaszy, D. M., Deputte, B., Cooper, E. J., Colbert-White, E. N., & Hémery, C. (2011). When and how well can human-socialized capuchins match actions demonstrated by a familiar human? *American Journal of Primatology, 73*(7), 643–654. doi:10.1002/ajp.20941

Fragaszy, D. M. & Perry, S. (2003). *The biology of traditions*. Cambridge, UK: Cambridge University Press.

Franks, N. R., & Richardson, T. (2006). Teaching in tandem-running ants. *Nature, 439,* 153. doi:10.1038/439153a

Fritz, J., Bisenberger, A., & Kotrschal, K. (2000). Stimulus enhancement in graylag geese: Socially mediated learning of an operant task. *Animal Behaviour, 59,* 1119–1125. doi:10.1006/anbe.2000.1424

Galef, B. G. (1992). The question of animal culture. *Human Nature, 3,* 157–178. doi:10.1007/BF02692251

Heine, S. J. (2011). *Cultural psychology* (2nd ed.). New York, NY: W. W. Norton.

Heyes, C. M., & Galef, B. G. (1996). *Social learning in animals: The roots of culture*. Cambridge, MA: Academic Press.

Heyes, C. M., Ray, E. D., Mitchell, C. J., & Nokes, T. (2000). Stimulus enhancement: Controls for social facilitation and local enhancement. *Learning and Motivation, 31,* 83–98. doi:10.1006/lmot.1999.1041

Hirata, S., Watanabe, K., & Kawai, M. (2001). "Sweet-potato washing" revisited. In T. Matsuzawa (Ed.), *Primate origins of human cognition and behavior* (pp. 487–508). Tokyo, JPN: Springer.

Holbrook, C. T., Barden, P. M., & Fewell, J. H. (2011). Division of labor increases with colony size in the harvester ant *Pogonomyrmex californicus*. *Behavioral Ecology, 22,* 960–966. doi:10.1093/beheco/arr075

Hoppitt, W. J. E., Brown, G. R., Kenal, R., Rendell, L., Thornton, A., Webster, M. M., & Laland, K. N. (2008). Lessons from animal teaching. *Trends in Ecology & Evolution, 23,* 486–493. doi:10.1016/j.tree.2008.05.008

Horner, V., & Whiten, A. (2005). Causal knowledge and imitation/emulation switching in chimpanzees (Pan troglodytes) and children (Homo sapiens). *Animal Cognition, 8*(3), 164–181. doi:10.1007/s10071-004-0239-6

Iacoboni, M. (2009). Imitation, empathy, and mirror neurons. *Annual Review of Psychology, 60,* 653–670. doi:10.1146/annurev.psych.60.110707.163604

Iacoboni, M., & Dapretto, M. (2006). The mirror neuron system and the consequences of its dysfunction. *Nature Reviews Neuroscience, 7,* 942–951. doi:10.1038/nrn2024

Ikkatai, Y., Watanabe, S., & Izawa, E. I. (2016). Reconciliation and third-party affiliation in pair-bond budgerigars (*Melopsittacus undulatus*). *Behaviour, 153,* 1173–1193. doi:10.1163/1568539X-00003388

Imanishi, K. (1952). The evolution of human nature (in Japanese). In K. Imanishi (Ed.), *Ningen* (pp. 36–94). Tokyo, JPN: Mainichi-shinbunsha.

Insel, T. R., & Shapiro, L. E. (1992). Oxytocin receptor distribution reflects social organization in monogamous and polygamous voles. *Proceedings of the National Academy of Sciences, 89*(13), 5981–5985. doi:10.1073/pnas.89.13.5981

Jaakkola, K., Rodriguez, M., & Guarino, E. (2010). Blindfolded imitation in a bottlenose dolphin (*Tursiops truncatus*). *International Journal of Comparative Psychology, 23*(4), 671–688. Retrieved from http://comparativepsychology.org/ijcp-2010-4/08.Jaakkola_etal_Final.pdf

Joly-Mascheroni, R. M., Senju, A., & Shepherd, A. J. (2008). Dogs catch human yawns. *Biology Letters, 4*, 446–448. doi:10.1098/rsbl.2008.0333

Kawai, M. (1965). Newly-acquired pre-cultural behavior of the natural troop of Japanese monkeys on Koshima islet. *Primates, 6*(1), 1–30. doi:10.1007/BF01794457

Kawamura, S. (1954). On a new type of feeding habit which developed in a group of wild Japanese monkeys (in Japanese). *Seibutsu-shinka, 2*, 11–13.

Kirchhofer, K. C., Zimmermann, F., Kaminski, J., & Tomasello, M. (2012). Dogs (*Canis familiaris*), but not chimpanzees (*Pan troglodytes*), understand imperative pointing. *PLoS One, 7*, e30913. doi:10.1371/journal.pone.0030913

Klinnert, M. D., Campos, J. J., Sorce, J. F., Emde, R. N., & Svejda, M. (2013). Emotions as behavior regulators: Social referencing in infancy. In R. Plutchik & H. Kellerman (Eds.), *Emotions in early development* (Vol. 2, pp. 57–86). New York, NY: Academic Press.

Knapska, E., & Meyza, K. Z. (2018). Introduction—Empathy beyond semantics. In E. Knapska & K. Z. Meyza (Eds.), *Neuronal correlates of empathy: From rodent to human* (pp. 1–6). London, UK: Elsevier.

Kopps, A. M., Ackermann, C. Y., Sherwin, W. B., Allen, S. J., Bejder, L., & Krützen, M. (2014). Cultural transmission of tool use combined with habitat specializations leads to fine-scale genetic structure in bottlenose dolphins. *Proceedings of the Royal Society B: Biological Sciences, 281*(1782), 20133245. doi:10.1098/rspb.2013.3245

Krützen, M., Mann, J., Heithaus, M. R., Connor, R. C., Bejder, L., & Sherwin, W. B. (2005). Cultural transmission of tool use in bottlenose dolphins. *Proceedings of the National Academy of Sciences, 102*, 8939–8943. doi:10.1073/pnas.0500232102

Kuczaj, S. A., Makecha, R., Trone, M., Paulos, R. D., & Ramos, J. A. (2006). Role of peers in cultural innovation and cultural transmission: Evidence from the play of dolphin calves. *International Journal of Comparative Psychology, 19*(2), 223–240. Retrieved from http://escholarship.org/uc/item/4pn1t50s.pdf

Laland, K. N., & Williams, K. (1998). Social transmission of maladaptive information in the guppy. *Behavioral Ecology, 9*, 493–499. doi:10.1093/beheco/9.5.493

LaMarca, K., Gevirtz, R., Lincoln, A. J., & Pineda, J. A. (2018). Facilitating neurofeedback in children with autism and intellectual impairments using TAGteach. *Journal of Autism and Developmental Disorders, 48*(6), 2090–2100. doi:10.1007/s10803-018-3466-4

Lamon, N., Neumann, C., Gruber, T., & Zuberbühler, K. (2017). Kin-based cultural transmission of tool use in wild chimpanzees. *Science Advances, 3*, e1602750. doi:10.1126/sciadv.1602750

Langford, D. J., Crager, S. E., Shehzad, Z., Smith, S. B., Sotocinal, S. G., Levenstadt, J. S., … Mogil, J. S. (2006). Social modulation of pain as evidence for empathy in mice. *Science, 312*, 1967–1970. doi:10.1126/science.1128322

Larsen, C. S. (2017). *Our origins: Discovering physical anthropology* (4th ed.). New York, NY: W. W. Norton.

Levy, I. M., Pryor, K. W., & McKeon, T. R. (2016). Is teaching simple surgical skills using an operant learning program more effective than teaching by demonstration? *Clinical Orthopaedics and Related Research, 474*(4), 945–955. doi:10.1007/s11999-015-4555-8

Lopez, J. C., & Lopez, D. (1985). Killer whales (*Orcinus orca*) of Patagonia, and their behavior of intentional stranding while hunting nearshore. *Journal of Mammalogy, 66*, 181–183. doi:10.2307/1380981

Mathis, A., Chivers, D. P., & Smith, R. J. F. (1996). Cultural transmission of predatory recognition in fishes: Intraspecific and interspecific learning. *Animal Behaviour, 51*, 185–201. doi:10.1006/anbe.1996.0016

McGrew, W. C. (2004). *The cultured chimpanzee: Reflections on cultural primatology*. Cambridge, UK: Cambridge University Press.

Merola, I., Lazzaroni, M., Marshall-Pescini, S., & Prato-Previde, E. (2015). Social referencing and cat–human communication. *Animal Cognition, 18*, 639–648. doi:10.1007/s10071-014-0832-2

Merola, I., Prato-Previde, E., & Marshall-Pescini, S. (2012). Social referencing in dog-owner dyads? *Animal Cognition, 15*, 175–185. doi:10.1007/s10071-011-0443-0

Meyza, K. Z., Bartal, I. B. A., Monfils, M. H., Panksepp, J. B., & Knapska, E. (2017). The roots of empathy: Through the lens of rodent models. *Neuroscience & Biobehavioral Reviews, 76*, 216–234. doi:10.1016/j.neubiorev.2016.10.028

Miller, H. C., Rayburn-Reeves, R., & Zentall, T. R. (2009). Imitation and emulation by dogs using a bidirectional control procedure. *Behavioural Processes, 80*, 109–114. doi:10.1016/j.beproc.2008.09.011

Miller, R., Logan, C. J., Lister, K., & Clayton, N. S. (2016). Eurasian jays do not copy the choices of conspecifics, but they do show evidence of stimulus enhancement. *PeerJ, 4*, e2746. doi:10.7717/peerj.2746

Mineka, S., Davidson, M., Cook, M., & Keir, R. (1984). Observational conditioning of snake fear in rhesus monkeys. *Journal of Abnormal Psychology, 93*, 355–372. doi:10.1037/0021-843X.93.4.355

Morimoto, Y., & Fujita, K. (2012). Capuchin monkeys (*Cebus apella*) use conspecifics' emotional expressions to evaluate emotional valence of objects. *Animal Cognition, 15*(3), 341–347. doi:10.1007/s10071-011-0458-6

Noad, M. J., Cato, D. H., Bryden, M. M., Jenner, M. N., & Jenner, K. C. (2000). Cultural revolution in whale songs. *Nature, 408*(6812), 537. doi:10.1038/35046199

Paukner, A., Suomi, S. J., Visalberghi, E., & Ferrari, P. F. (2009). Capuchin monkeys display affiliation toward humans who imitate them. *Science, 325*, 880–883. doi:10.1126/science.1176269

Persicke, A., Jackson, M., & Adams, A. N. (2014). Brief report: An evaluation of TAGteach components to decrease toe-walking in a 4-year-old child with autism. *Journal of Autism and Developmental Disorders, 44*(4), 965–968. doi:10.1007/s10803-013-1934-4

Pitman, R. L., Deecke, V. B., Gabriele, C. M., Srinivasan, M., Black, N., Denkinger, J., ... Ternullo, R. (2017). Humpback whales interfering when mammal-eating killer whales attack other species: Mobbing behavior and interspecific altruism? *Marine Mammal Science, 33*, 7–58. doi:10.1111/mms.12343

Platek, S. M., Critton, S. R., Myers, T. E., & Gallup, G. G. (2003). Contagious yawning: The role of self-awareness and mental state attribution. *Cognitive Brain Research, 17*, 223–227. doi:10.1016/S0926-6410(03)00109-5

Pope, S. M., Taglialatela, J. P., Skiba, S. A., & Hopkins, W. D. (2017). Changes in frontoparietotemporal connectivity following Do-As-I-Do imitation training in chimpanzees (*Pan troglodytes*). *Journal of Cognitive Neuroscience, 30*, 421–431. doi:10.1162/jocn_a_01217

Premack, D. (1959). Toward empirical behavior laws: I. Positive reinforcement. *Psychological Review, 66*(4), 219–233. doi:10.1037/h0040891

Pryor, K. W. (1999). *Don't shoot the dog: The new art of teaching and training.* New York, NY: Bantam Books.

Quervel-Chaumette, M., Faerber, V., Faragó, T., Marshall-Pescini, S., & Range, F. (2016). Investigating empathy-like responding to conspecifics' distress in pet dogs. *PLoS One, 11*(4), e0152920. doi:10.1371/journal.pone.0152920

Quinn, M. J., Miltenberger, R. G., & Fogel, V. A. (2015). Using tagteach to improve the proficiency of dance movements. *Journal of Applied Behavior Analysis, 48*(1), 11–24. doi:10.1002/jaba.191

Reimert, I., Bolhuis, J. E., Kemp, B., & Rodenburg, T. B. (2015). Emotions on the loose: Emotional contagion and the role of oxytocin in pigs. *Animal Cognition, 18*(2), 517–532. doi:10.1007/s10071-014-0820-6

Rendell, L. E., & Whitehead, H. (2001). Culture in whales and dolphins. *Behavioral and Brain Sciences, 24*(2), 309–382. doi:10.1017/S0140525X0100396X

Rizzolatti, G., & Craighero, L. (2004). The mirror-neuron system. *Annual Review of Neuroscience, 27*, 169–192. doi:10.1146/annurev.neuro.27.070203.144230

Russell, C. L., Bard, K. A., & Adamson, L. B. (1997). Social referencing by young chimpanzees (*Pan troglodytes*). *Journal of Comparative Psychology, 111*, 185. doi:10.1037/0735-7036.111.2.185

Sigler, E. A., & Aamidor, S. (2005). From positive reinforcement to positive behaviors: An everyday guide for the practitioner. *Early Childhood Education Journal, 32*(4), 249–253. doi:10.1007/s10643-004-0753-9

Silberberg, A., Allouch, C., Sandfort, S., Kearns, D., Karpel, H., & Slotnick, B. (2014). Desire for social contact, not empathy, may explain "rescue" behavior in rats. *Animal Cognition, 17*, 609–618. doi:10.1007/s10071-013-0692-1

Shreve, K. R. V., & Udell, M. A. (2015). What's inside your cat's head? A review of cat (*Felis silvestris catus*) cognition research past, present and future. *Animal Cognition, 18*, 1195–1206. doi:10.1007/s10071-015-0897-6

Suchak, M., Watzek, J., Quarles, L. F., & de Waal, F. B. (2018). Novice chimpanzees cooperate successfully in the presence of experts, but may have limited understanding of the task. *Animal Cognition, 21*, 87–98. doi:10.1007/s10071-017-1142-2

TAGteach International. (n.d.). Retrieved June 13, 2017, from http://www.tagteach.com/

TAGteach International. (2004). *Teaching with acoustical guidance: TAGteach*. Retrieved from http://www.tagteach.com/

Tomasello, M. (1996). Do apes ape? In C. M. Heyes & B. G. Galef Jr. (Eds.), *Social learning in animals: The roots of culture* (pp. 319–346). New York, NY: Academic Press.

Ueno, A., & Matsuzawa, T. (2005). Response to novel food in infant chimpanzees: Do infants refer to mothers before ingesting food on their own? *Behavioural Processes, 68*, 85–90. doi:10.1016/j.beproc.2004.09.002

van de Waal, E., Borgeaud, C., & Whiten, A. (2013). Potent social learning and conformity shape a wild primate's foraging decisions. *Science, 340*, 483–485. doi:10.1126/science.1232769

van Lawick-Goodall, J. (1968). The behavior of free-living chimpanzees in the Gombe Stream Reserve. *Animal Behaviour, 1*, 161–311. doi:10.1016/S0066-1856(68)80003-2

Walden, T. A., & Ogan, T. A. (1988). The development of social referencing. *Child Development, 59*, 1230–1240. doi:10.2307/1130486

Walle, E. A., Reschke, P. J., & Knothe, J. M. (2017). Social referencing: Defining and delineating a basic process of emotion. *Emotion Review, 9*, 245–252. doi:10.1177/1754073916669594

Webster, S. J., & Fiorito, G. (2001). Socially guided behaviour in non-insect invertebrates. *Animal Cognition, 4*, 69–79. doi:10.1007/s100710100108

Webb, C. E., Romero, T., Franks, B., & de Waal, F. B. M. (2017). Long-term consistency in chimpanzee consolation behaviour reflects empathetic personalities. *Nature Communications, 8*, 292. doi:10.1038/s41467-017-00360-7

Wechkin, S., Masserman, J. H., & Terris, W. (1964). Shock to a conspecific as an aversive stimulus. *Psychonomic Science, 1*, 47–48. doi:10.3758/BF03342783

Wolfle, D. L., & Wolfle, H. M. (1939). The development of cooperative behavior in monkeys and young children. *The Pedological Seminary and Journal of Genetic Psychology, 55*, 137–175. doi:10.1080/08856559.1939.10533188

Yoon, J. M. D., & Tennie, C. (2010). Contagious yawning: A reflection of empathy, mimicry, or contagion? *Animal Behaviour, 79*, e1–e3. doi:10.1016/j.anbehav.2010.02.011

Cognitive Flexibility in Animals

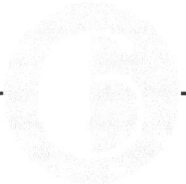

If you've ever been faced with a deer in your headlights, you may have thought, "Why won't that *stupid* deer move out of the way?" Part of the reason is temporary blindness from the bright lights, but the rest has to do with how deer respond to threats. According to the behavioral systems approach to understanding organisms, when a deer's threat system is activated, the options are to freeze, flee, or fight (e.g., Timberlake & Lucas, 1989). For prey animals like deer (and apparently even humans to some extent; Azevedo & Van Sluys, 2005), the system activates freeze. Interpreting the deer's behavior this way, rather than assuming it's making a stupid decision, allows you the opportunity to appreciate millions of years of evolution in action. The deer is doing exactly what it's supposed to do when it encounters a threat.

Wait, so does this mean Descartes was right all along? Are animals basically just machines—systems responding reflexively with a limited set of behaviors? Yes and no. Animals experience a constant tension between fixed and flexible responses to their environment. Under some circumstances, it's actually good to be like a machine.

In others, like in the deer's case, being able to override the reflex to freeze when a car is coming is more beneficial.

When we think about cognitively flexible animals, we tend to say that they can (1) quickly modify their behavior (i.e., go off script) in order to respond to unexpected variations in their environment and (2) do so even in circumstances that are unrelated to survival, such as play behavior (Bond, Kamil, & Balda, 2007; Burghardt, 2005). Not surprisingly, cognitive flexibility varies with features of an animal's natural history. That is to say, species are uniquely suited to behave and adapt their responding in ways that are relevant to the physical and social environment in which they evolved. A great example of this is illustrated by ring-tailed lemurs (*Lemur catta*), a species of primate that can only be found on the African island country of Madagascar. Ring-tailed lemurs have a very rich and complex social organization, meaning recognizing individuals and their respective positioning within the hierarchy is important. As a result, they perform exceptionally well on tests that assess transitive inference, that is, if A is greater than B and B is greater than C, A must also be greater than C. It turns out that the lemur's natural social organization makes them particularly skilled at solving these kinds of relational problems in the laboratory (MacLean, Merritt, & Brannon, 2008). The moral of the story here is simply that lemurs may never solve calculus problems like humans can, but they also do not encounter problems in their lives that would necessitate understanding it.

INSTINCTS

The phrase *nature versus nurture* was a famous debate among psychologists about whether behavior is innately driven or learned. We now know that behavior and cognitive processes are guided by a dynamic interaction between biology and the environment. For example, being biologically human is not enough for a neurotypically developing child to acquire language. There must be a learning component that comes from repeatedly experiencing and practicing language during a critical period of time in development.

Even still, it is not that simple! Research in the new field of epigenetics indicates that our environment can also change how our genes function and that those changes can be passed down to offspring. Further, instinctual behaviors may have evolved from

CHAPTER 6 COGNITIVE FLEXIBILITY IN ANIMALS

behaviors that were originally learned thousands of generations before (Robinson & Barron, 2017). This would mean that a complex courtship display that now seems reflexive and instinctual could be traced back, perhaps, to one male's innovative deviation from the boring courtship display his competition was doing. This deviation may have been more attractive to females than the original display, so he passed on lots of his genes. Fast-forward to today, and we look at the dance and write it off as being innate without considering how it might have originated.

Referring to a behavior as instinct-driven seems to suck the cognition and intelligence out of it and replace it with automaticity and mindlessness. We endorse a less judgmental, more open-minded approach to thinking about instinctual behavior. Survival requires constant vigilance in order to detect and respond appropriately to threats, food, mates, and other environmental stimuli. By hardwiring some responses as instincts, organisms can devote more time to other things, such as more complex and flexible cognition. Along these same lines, learning a behavior is much easier when an organism is biologically predisposed to certain behaviors. For example, dogs appear to be biologically predisposed to attend to and follow human pointing cues because of their coevolution with humans (Hare, Brown, Williamson, & Tomasello, 2002). By being "preprogrammed" with pointing sensitivity, dogs start at an advantage for understanding human gestural communication, thus increasing the efficiency and sophistication of their interactions with us.

Instincts are also especially helpful when it comes to bypassing the hazards of trial-and-error learning. Imagine a mouse having to learn to avoid the scent of a cat by experiencing firsthand the damage it can do! Likewise, rather than having to learn to avoid putrid food because it can make them sick, many animals have a built-in disgust system that is triggered by the presence of rot. This gives the animal a heads-up that it should proceed with caution before eating.

For an organism that lives in a relatively unchanging environment, where rot, the scent of cats, and the preferences of females rarely change or change very slowly, there's not a lot of need to think on your feet, so to speak, so offloading cognition to instincts makes sense. But what about animals who live in unpredictable environments? In that case, it's far more advantageous to be able to go off script and respond flexibly. Indeed, living in a changing environment predicts greater cognitive flexibility, presumably because individuals will have to effectively handle a greater diversity of predators, competitors, resources, and experiences (Bond, Diamond, & Bond, 2003; Bond et al., 2007; Japyassú & Laland, 2017). Along the same lines, Japyassú and Laland (2017) also speculate animals with generalist diets (i.e., eat lots of different foods) should show greater cognitive flexibility than those with specialist diets. Parrots are a great example of another kind of cognitively flexible generalist. Parrots learn their communication systems from members of their flock. As a testament to their vocal generalist nature, young parrots raised by parents of an entirely different parrot species will adapt and learn to use the vocalizations of their foster parents, thereby ensuring they are cared for and included in the group (Rowley & Chapman, 1986). Even when their foster parents are humans, parrots' predisposition to learn to use the communication system they're exposed to allows them to communicate using speech with their new "flock" (Colbert-White, Covington, & Fragaszy, 2011; see work by Irene Pepperberg as well).

While parrots are a great example of flexibility in the face of unpredictable environments, another bird, the cuckoo (*Cuculus canorus*), teaches us an opposite lesson. The cuckoo avoids parental responsibility by laying its eggs in the nests of other species. Because cuckoo chicks grow larger than their host parents' chicks, the cuckoo chicks' gaping mouth is bigger and stands out more (Tanaka, Morimoto, Stevens, & Ueda, 2011). The bigger, brighter mouth sends the host parents' feeding instinct into overdrive, and the cuckoo chicks eventually eliminate their competition. The cuckoo shows us that when it comes to instincts, offloading cognition to a script can be helpful, but a bit of cognitive flexibility always helps.

Another place where instincts can cause complications is in the learning process. According to behaviorists, any behavior could be learned by any organism. Two of B. F. Skinner's former students proved him wrong. In 1961, Marian Breland Bailey and Keller Breland wrote "The Misbehavior of Organisms." The article chronicles "egregious failures" (p. 683) of some of the animals they had been hired to train for television. In one case, a pig was trained to put coins into a piggy bank, but once the pig reached peak performance, the behavior curiously began to break down. The pig began dropping the coins on the floor, rooting at them with its nose, and tossing them in the air. This behavior was puzzling until the Brelands realized this was exactly how pigs forage and interact with food. The coins had taken on too great of an association with the food rewards, triggering the pig's natural foraging system. The Brelands called this phenomenon instinctive drift, and in four short pages, they rejected the decades-old assumption that learning was a one-size-fits-all process, where every species is essentially a blank slate and behaviors are equally trainable.

The Brelands were among the first psychologists to really explore the importance of thinking about natural history and the role of instincts with respect to animal learning and behavior. Now, animal cognition researchers encourage these sorts of consideration when it comes to developing hypotheses and methodologies as well as interpreting findings.

PLANNING AND FORETHOUGHT

As a child, you may have heard the story of the grasshopper and the ant as a lesson in not putting off until tomorrow what you can do today. Can ants, or any other animal for that matter, plan for the future? One perspective is that animals seem so carefree because they only live in the here and now (Robert Sapolsky's *Why Zebras Don't Get Ulcers* is an interesting read that plays on this idea). On the contrary, there is mounting evidence that at least some other species modify their behavior in anticipation of a future reality. Terms associated with this ability are *planning, forethought, mental time travel,* and *prospective memory* (i.e., "remembering to remember," such as remembering you need to drop off a letter at the post office on the way home when you get off work in 6 hours). We'll use the first three terms interchangeably while recognizing a distinction between them and *prospective memory*.

The cognitive processes necessary to project into the future epitomize a highly flexible mind. Nonetheless, evolutionarily speaking, mental time travel may not be useful for all animals. Instead, it may be more useful to have sophisticated systems in place to respond to the present. For example, it's not always helpful to worry about rivals, mates, danger, or food sources that existed yesterday or may exist tomorrow, especially if that worrying impedes the individual's ability to effectively respond to the present. This might sound familiar, as we often refer to the debilitating effects of worrying about the past and the future as anxiety, which can severely impair daily functioning.

For other species, the reduced costs associated with planning in response to future expectations can save an animal time and resources. One dolphin demonstrated this efficiency in a study that required him to find weights throughout his enclosure and deposit them into a spring-loaded machine that released a fish when enough weight was added. When the weights were placed near the machine, Bob the dolphin would go one by one, retrieving a weight and dropping it into the machine until the fish was released. However, when the weights were placed up to 45 meters from the machine, it only took about 10 times of swimming back and forth all over the enclosure before Bob started collecting the weights in groups of up to five at a time before returning to the machine to deposit them. The fact that he did not do this behavior when the weights were close to the machine indicated to the researchers that Bob was efficiently planning for the future (Kuczaj, Gory, & Xitco, 2009).

In the wild, field researchers assess animals' group movement patterns to evaluate the extent to which their behavior illustrates this kind of efficiency. Using GPS (global positioning system) technology, primates' travel routes can be tracked as the animals travel to food sites. A group leader that is planning ahead will construct an efficient travel path that reduces energetic costs, while a leader that is responding in the moment from site to site may lead the group to backtrack or zigzag. Efficiency in routes is known as the traveling salesman problem and has been studied in the field as well as the laboratory (e.g., Howard & Fragaszy, 2014; Valero & Byrne, 2007). One research team was surprised to discover that male orangutans announce the direction they will be traveling anywhere from a few hours to up to a day in advance (van Schaik, Damerius, & Isler, 2013). Because orangutans are solitary animals, by broadcasting their future plans, rival males would know which areas to avoid and females would know where to find a potential mate.

Enter the cleaner wrasse (*Labroides dimidiatus*). This small fish excels in tests designed to measure food source decisions and planning for the future. Initial tasks involved multiple stations of preferred food and one station of a nonpreferred food. The nonpreferred food was always refilled after eating. The preferred food stations were refilled after particular amounts of time had elapsed (5, 10, or 15 minutes). Here, the wrasses planned their feeding to maximize amounts of the preferred food based on how much time had elapsed since they had fed at each station (Salwiczek & Bshary, 2011). In a second set of experiments, the wrasses were made to choose between two different colored plates of food. One of the plates would always be refilled if it was chosen ("permanent"), and the other would not ("ephemeral"). If the permanent plate was chosen, the ephemeral one would be removed. An animal who is able to think about the future should always choose the ephemeral plate, and an animal who cannot should choose between the two plates at random. Why? Because even though the all-you-can-eat permanent plate might seem tempting, by choosing the ephemeral plate first and *then* shifting to the permanent plate, the subject maximizes the amount of food it receives. The task is also complicated by the fact that the plate chosen at the start of the trial will always yield the same amount of food, so unless you're able to think two steps ahead, either of the plates will do. Not only did the wrasses maximize their reward, but they also outperformed chimpanzees, capuchins, and orangutans, calling into question why our closest kin were not able to blow the wrasses out of the water (Salwiczek et al., 2012).

A final champion of future planning comes from the western scrub jay. These birds are well known for their abilities to cache or store food for the future, and some of the first studies of memory were done with these birds. In these experiments, the birds were allowed to cache foods that spoiled fairly quickly (mealworms) and foods that did not (peanuts). If a shorter time between caching and retrieval elapsed, the birds would retrieve the "perishables" first—just like you bring the bag with the ice cream in it into the house first after you go shopping. However, if a longer amount of time passed, the birds would head straight for the peanuts, not bothering to retrieve the already-spoiled mealworms (Clayton & Dickinson, 1998). In addition, the same research group showed that jays will store food in locations where they're more likely to be hungry in the future and in locations where food is less likely to be readily accessible (Raby, Alexis, Dickinson, & Clayton, 2007). They'll also selectively cache preferred foods based on a future expectation. Let's say your favorite food was pizza. And I let you eat pizza until you're bursting at the seams. Then, I gave you the choice between taking home *more* pizza or a grilled cheese. In that moment, pizza does not seem very appetizing, which means if you were not able to plan for the future, you'd probably take the grilled cheese. But since you can project into tomorrow and know you'd be ready for pizza again, you choose the pizza. This is exactly what the jays did when presented with the same scenario involving pine nuts and kibble (Correia, Dickinson, & Clayton, 2007).

In addition to the species we've just highlighted, there's a surprising amount of diversity in the species that appear able to plan ahead and/or remember to remember, including crows, rats, dolphins, killer whales, and octopuses[1] (see Crystal, 2013; Klump, Sugasawa, St Clair, & Rutz, 2015; Kuczaj & Walker, 2006; Kuczaj, Stan, Xitco, & Gory, 2010). Yes, octopuses. During a 9-year longitudinal study, Finn, Tregenza, and Norman (2009) noticed many veined octopuses (*Amphioctopus marginatus*) used coconut shells as hiding places. While this is already a remarkable example of tool use by an invertebrate, the octopuses carried the shells around with them by tucking them under their body and walking along the ocean floor on their tiptoes. Even when Finn et al. scared the octopuses away from the shells, they would always come back later to retrieve them.[2] The authors noted that traveling with the shells actually leaves them more vulnerable to predation and requires a great deal of energy expenditure, so the octopuses may recognize the future benefits of carrying their awkward coconut armor outweigh the costs of being eaten, indicating true cognitive flexibility.

Here's thought to consider: Why do people grab an umbrella before heading out into the rain? Likely, they have connected being caught out in the rain with being cold and wet and umbrellas with escaping those negative sensations. Can grabbing an umbrella on the way out the door only be explained by complex future planning? If so, we would need to concede that amoebas (*Amoeba* spp.) can plan for the future because they, too, respond in the present in anticipation of learned future outcomes. Amoebas are single-celled organisms, and if given a proper cue, they learn to behave in seeming anticipation of some other stimulus. In the case of De la Fuente et al.'s (2018) study, the cue was an electric field that signaled food would be present if the amoeba moved to a particular location. Amazingly enough, their "memory" for this learned behavior lasted up to 4 hours. Would you be comfortable saying amoebas can plan for the future? If not, then we need to be just as critical of the fact that other species deemed cognitively flexible might also be operating under the same basic learning principles.

---

1. *Octopuses* is the correct grammar. Or octopodes, if you prefer.
2. "Hi, hon, what did you do at work today?" "Scared the heck out of a bunch of octopuses." "That's nice dear."

PROBLEM SOLVING

In one of his fables, Aesop described a clever crow dropping stones into a pitcher until the water level rose high enough for him to drink. Aesop was really an animal cognition researcher at heart, as this fable has been successfully translated into a foraging-based task with a variety of species, including orangutans (Mendes, Hanus, & Call, 2007), crows (Jelbert, Taylor, Cheke, Clayton, & Gray, 2014), and even wild-caught raccoons (*Procyon lotor*; Stanton, Davis, Johnson, Gilbert, & Benson-Amram, 2017). To be successful, the animal has to overcome functional fixedness, the tendency to get hung up on the function of an object and be unable to see it for a new purpose. When I use a butter knife because I do not have a real screwdriver, I have overcome the fixed mental state of viewing a butter knife as only for cutting food. The crow's ability to see stones as weight that can displace water, and not just part of its environment, has been said to rival a 5- to 7-year-old human child's physical understanding of the world (Jelbert et al., 2014).

You may have heard the terms *insight* or *aha moment* to describe the point in time when the clouds part and the solution to a problem becomes clear. What you may not know is that shortly after insight learning was first empirically described in humans, it was documented in another primate. At a research station in the Canary Islands, Wolfgang Köhler (1887–1968) worked with many different species, but it was his chimpanzee problem-solving studies for which he is best known. His book, *The Mentality of Apes* (1925), chronicles a series of studies he conducted to understand how we differ from our closest kin with respect to problem solving. At the end of the day, Köhler argued there was not much difference and that humans had long underestimated other species.

Two of Köhler's classic problem-solving tasks were the stick problem and the box problem. In both, fruit was placed just out of reach of a hungry chimpanzee. The subject had to either connect two sticks together to make a longer stick or stack a few boxes and climb on them in order to reach the fruit. After some hours of failed attempts, Köhler noted that the subjects would pause, appear to solve the problem in their heads, and then move to enact that solution in a very intentional way. According to Köhler, the pattern of behavior mirrored what was observed in humans solving similar problems and thus offered evidence for insight learning in other species.

Among both human and nonhuman psychologists, Köhler is easily one of the most famous animal cognition researchers, and his methods have been replicated with multiple species. Grizzly bears (*Ursus arctos*) tend to make people think about aggression more than patient problem solving. Nonetheless, just like Köhler's chimpanzees, grizzly bears can use logs as tools to reach doughnuts—their reward of choice—suspended from a rope (Waroff, Fanucchi, Robbins, & Nelson, 2017). Others have opted to follow in Köhler's footsteps in spirit by branching out to create new problem-solving tasks. From hyenas (*Crocuta crocuta*) figuring out how to open puzzle boxes (Benson-Amram, Weldele, & Holekamp, 2013) to archerfish (*Toxotes jaculatrix*) compensating for visual distortion when they shoot targets outside their tank with a stream of water (Schuster, Rossel, Schmidtmann, Jäger, & Poralla, 2004), every major group of animals has been shown to solve novel problems—offering compelling accounts of impressive feats of cognitive flexibility throughout the animal kingdom. As you might imagine, it would be exhausting both for us as writers and for you as the reader to include all of the species that have been tested over the years. Thus, like the previous section, we present some of our favorite examples framed around common methodologies and relevant considerations.

An extraction task is any kind of task in which an animal must work to remove a food item that is contained behind or within a barrier. One common extraction task is the string pull task, in which an animal is presented with food that has been tied to the bottom of a long string suspended from a perch. The food is unreachable from the air, usually because it hangs down into a clear plastic tube. To extract the food, the subject must recognize that the string can be pulled up. The challenge for the animal is to figure out how to accomplish this. Usually it involves the subject repeatedly pulling up on a segment of the string and stepping on it to hold it in place. According to some, this innovative solution represents a form of insightful problem solving akin to Köhler's apes (e.g., Heinrich, 1995). Stereotypically intelligent birds like parrots, social birds of prey, and corvids tend to perform the best on this task (Colbert-White, McCord, Sharpe, & Fragaszy, 2013; Heinrich, 1995; Pepperberg, 2004). On the other hand, after less than 10 minutes of buzzing, inspecting, failed reaching, and testing the apparatus, 2 out of 25 bumblebees spontaneously figured out how to pull a horizontal string to reach a tiny bowl of sugar water (Alem et al., 2016). Naive bumblebees who watched these successful problem solvers were also faster to solve

CHAPTER 6   COGNITIVE FLEXIBILITY IN ANIMALS

the problem than nonwatchers, demonstrating how a novel behavior could be passed down to members of a social group via observation!

For both societal and methodological reasons, exploration of reptiles' problem-solving abilities was minimal until recently. At one point, reptiles were written off as unintelligent and slow to learn, only to discover later that the mammal-friendly laboratory temperatures in which these cold-blooded animals were tested were actually preventing them from performing their best (Wilkinson & Glass, 2018). Try focusing on solving a Sudoku puzzle while sitting outside in your pajamas in December. You might not be at the top of your game either! Though reptiles diverged from birds and mammals nearly 300 million years ago, their problem-solving abilities are quite impressive, offering a window into the evolution of cognitive flexibility in the animal kingdom. In one study, researchers designed a problem that required a small tropical lizard to completely shift its hunting style in order to succeed. Over millions of years of evolution, the emerald anole (*Anolis evermanni*) became a predator who attacked unsuspecting prey from above. Leal and Powell (2012) presented the lizards with a tasty grub under a bottle cap–shaped lid. While some of their subjects kept trying to pounce down onto the lid to get to the grub, a few of them developed novel behaviors like approaching the lid from the side and flipping it up with their snout—behaviors that were a clear deviation from their instinctual hunting script.

Logically speaking, organisms that live in a complex world should need complex brains in order to keep up with their constantly changing environment. What if we told you a brain smaller than the head of a pin, containing fewer than 1 million neurons (humans have 100 billion), can produce behavior like flexible route detouring (Harland & Jackson, 2004)? Jumping spiders embody the saying "Good things come in small packages." These tiny arachnids are ambush predators, rather than web builders. Their specialized visual system includes four pairs of eyes that see in full-color vision, with nearly a 360° view, and visual acuity that is better than some mammals' (Menda, Shamble, Nitzany, Golden, & Hoy, 2014). Using vision as their guide, jumping spiders can lock in on prey from nearly a foot away, stalk them, and then pounce.

For decades, researchers have studied jumping spiders' ability to create detours, or shortcuts, to their prey in order to be more efficient. The environment of a tiny jumping spider can include rocks, leaves, and other obstructions. For a hungry animal, the prey-seeking drive to keep moving forward toward prey can lead to inefficiency or exposure. Jumping spiders, instead, somehow seem to create mental maps of their location with respect to their potential prey, sometimes making reversed-route detours in which they move away from their prey in order to eventually get closer to it. This overriding of the instinct to always move toward prey, mapping locations, and detouring around obstructions represents a combination of flexible problem-solving abilities, which some say also approaches planning or forethought (Tarsitano & Jackson, 1994).

One of the biggest criticisms of attributing complex problem-solving behavior to higher order cognitive abilities is Morgan's Canon. Put simply, when someone untangles a garden hose, it's much more likely that they're using basic trial-and-error learning—moving the hose one way, assessing the outcome, and responding again—as opposed to enacting some sophisticated, complex detangling sequence that they first created in their head. Indeed, in compiling the literature for this section, some of the authors were hesitant to describe their subjects' problem solving as the kind of insight learning Köhler attributed to his chimpanzees. For example, in the string-pull problem, a parrot might pick up the string a tiny bit while exploring the problem and inadvertently reinforce themselves because picking up the string physically brought the food reward closer, even for a second. According to behaviorist Edward Thorndike's (1898) law of effect, those behaviors that have a pleasant outcome will be repeated. So, when the string is accidentally lifted and the food gets closer, the parrot does it again, and again, until the food is reached. Thus, what looks like one preplanned, highly intelligent problem-solving act might, in reality, be a series of isolated behavior-reward chains that are based on simple trial and error.

Like with string pulling, we might find it extraordinary that a grizzly bear seemingly *knows* to stack logs to reach a doughnut. Others, like behaviorists Epstein, Kirshnit, Lanza, and Rubin (1984), showed that with some pretraining, a pigeon presented with a tiny box would "insightfully" move it, step up onto it, and peck at a tiny plastic banana (an homage to Köhler's chimpanzees?) suspended above the bird's head. While the pigeons were never explicitly trained to do this whole sequence of behaviors in order, they had been separately rewarded for moving the box around the enclosure and standing on top of the box to peck the banana. That the pigeon observed the problem, put the steps together, and solved it is still impressive and deserves attention, but it does beg the question of whether insightful problem solving in animals may not be as preplanned or advanced as we think it is.

Taking it a step further, maybe the same could be said for insightful problem solving in humans. After all, human participants in Gick and Holyoak's (1980) experiment who had previously read a story about a castle being toppled by attacking it from multiple sides were significantly more likely to solve the following problem later via "insight": *New laser technology can destroy a cancerous tumor without surgery; however, if the laser is set on full power, it will damage healthy tissue around the tumor, and if it is set on lower power, it will be safe for healthy tissue but will not destroy the tumor. How can you destroy the tumor with the laser technology?* The solution is to use multiple lasers at lower intensity all aimed at the tumor, just like in the castle siege story.

Those who study animal problem solving must be highly attentive to concerns such as those stated earlier when they develop their methods and interpret their findings. The behaviors leading up to actually solving the problem are often just as interesting, if not more so, than the act of actually solving it. Behaviors such as cocking the head with gaze directed at the problem, circling the problem to get a better look or new angle, previous reinforcement or learning history (such as the castle–tumor problem), taking a break from the problem for a certain amount of time, and even the latency (i.e., time it takes until a behavior occurs) to begin solving the problem all can help better assess intentionality and, presumably, what might make a behavior an example of flexible problem solving rather than trial-and-error conditioning.

PLAY BEHAVIOR

My (AK) use of social media boils down to approximately three things: funny stories, pictures of my kids, and people sharing videos of their pets being adorable—that is, playing. Your social media does too—admit it.

We may recognize play when we see it—in either humans or animals—but for the study of animal cognition, a consistent definition or set of criteria is needed. One of the most well known was published in 2005 by animal researcher Gordon Burghardt. Burghardt requires five criteria for behavior to be classified as play:

(1) It must be nonfunctional for the environment in which it takes place. For example, small children who "play house" are not actually functioning as a domestic unit, nor is there any real threat when animals engage in play fighting.
(2) It must be voluntary and self-rewarding. Homework is not play. But the science kit Aunt Cindy got you for your 12th birthday is, particularly if it involves blowing things up.[3] It seems my dog plays with a squeaky toy because he finds the squeaks rewarding (and if not, he definitely finds it rewarding to dig the squeaker out, "kill" it, and leave fluff all over the house).
(3) It must be significantly different from other types of behavior in the species. There is nothing in the natural behavioral repertoire of humans that resembles tennis or Twenty Questions. Likewise, the kinds of chasing games played by the variety of animals living in my home do not generally end with anyone actually getting eaten.
(4) It must occur repeatedly over the species' life span—although the amount of time spent might change. Grown-ups play too, right? So does my 14-year-old dog we call "Old Man."
(5) It must occur in nonthreatening situations. Anyone who picks up a baseball bat while being followed down a dark alley is not looking to start a game. Likewise, young cubs wrestling over a piece of meat will both eventually get to eat in the end, so there's no real competition.

Though the definition may hold, what exactly play looks like across species will differ because it directly reflects a species' natural history, again illustrating the theme throughout this book of a connection between animals' behavior and their ecology (i.e., how animals interact with their environment). For example, bird species that are well known for their ability to solve problems and use tools—crows and parrots—seem more likely to play by manipulating objects in novel ways and in novel situations, for example, stacking square objects or rings. However, bird species that are known for their expertise at caching and storing food, such as jays, are more often observed playing with objects by placing them into holds or

---

3. That falls under the "rewarding" portion of the definition.

other small areas, for example, inserting a ball into a tube (Auersperg et al., 2015). This type of practicing and modeling of creative behavior is very reminiscent of the creative modeling and accumulation of expertise found while studying creativity in humans (Kaufman & Kaufman, 2007; Yi, Plucker, & Guo, 2015).

Amounts of play also differ by species. Current ideas for why have to do with energy expenditure. You cannot play if you're always exhausted from outrunning predators, foraging, defending your territory, or looking for water. Think about it, you'd be much more likely to play too if you were not working, caring for children, cooking, paying bills, or doing homework. Habitat can even matter, too. For example, because water is an easier medium in which to travel (it uses less energy), aquatic animals may be more likely to engage in play behaviors as they have "extra" energy resources to expend (Burghardt, 2014; Kuczaj & Eskelinen, 2014a).

Species that delay reproduction (relative to the rest of the life cycle) and pair bond, like parrots and some mammals, are more likely to play as well (Bond et al., 2003). Again, this makes sense as play would serve to strengthen bonds between mates. There also seems to be some correlation between brain size and play—bigger brained animals tend to play more. Presumably, larger brains allow for efficiency in daily tasks, which leaves time and energy for play (Iwaniuk, Nelson, & Pellis, 2001). Interestingly, this correlation only holds when comparing types of animals, for example, comparing carnivores to marsupials or rodents to carnivores. Comparing within a group (e.g., bears, cats, and wolves, which are all carnivores) does not show the same relationship (Iwaniuk et al., 2001). There also seems to be a correlation between the amount of play, with whom it occurs, and specific aspects of species social life. Tonkean macaques (*Macaca tonkeana*), who live in very cohesive, cooperative societies, are much more likely to play (and in particular, play between adults) than Japanese macaques (*Macaca fuscata*), who maintain a more aggressive and individualistic group (Ciani, Dall'Olio, Stanyon, & Palagi, 2012).

In addition to external features like habitat, environmental pressures, and social group, it seems there is at least some small genetic association inherent in a playful disposition. Evidence for this comes from a famous longitudinal study that bred wild foxes for "tameness" over 45 generations. These more domesticated, people-friendly foxes began to change physically, looking more like today's dogs, including floppy ears and short, curling tails. Most importantly, however, by breeding for tameness, Belyayev also discovered the foxes were more apt to play than their predecessors had been (Trut, 1999).

It makes sense that primates, cetaceans, and parrots are obvious candidates for play given their inclusion in almost every other section in this book, but do you know who also plays? Frogs! Dart frogs (*Dendrobatidae* spp.) engage in wrestling matches and tadpoles have been seen riding the bubble stream of an aerator to the top of the tank over and over (Burghardt, 2015). It's the "over and over" part that really connects to Burghardt's definition of being functionally irrelevant, intentional, and potentially self-rewarding. Reptiles, particularly lizards, often play in a dog-like fashion, tugging on toys, albeit very slowly. They will also explore boxes, containers, or other novel objects. Crocodiles (*Crocodylinae* spp.) and turtles (*Testudines* spp.) have been recorded interacting playfully with balls and hoses as well (Burghardt, 2015; Burghardt, Ward, & Rosscoe, 1996). Cephalopods like the octopus are notoriously intelligent and have been recorded doing things such as squirting caretakers with water and capturing and releasing food items despite being satiated (Burghardt, 2015; Zylinski, 2015).

Play serves many functions in both humans and nonhumans, such as teaching appropriate behaviors and skills. The "mouthing" your pet does when it plays with you takes time to master—many pet owners can remember a few times when their pet did not quite have the hang of it and bit down a little too hard. Play also hones survival skills (Kuczaj & Eskelinen, 2014b). Youngsters can attack and grapple with each other, which builds motor memory, and young predators learn to stalk and "kill" their prey by stalking each other (Caro, 1995). This is the reason we do things like practice sports as well—so that when you're in the middle of the game, your muscles will automatically remember how to dribble a basketball or catch a football without your brain having to think too much about it.

Play can also serve as a peacekeeper in societies where maintaining dominance hierarchies is vital to group cohesion. A simulated play "fight" can still have a winner and a loser and has the advantage of reduced bloodshed (Bauer & Smuts, 2007). Further, young members in the group who try to solicit play from others get a chance to learn their place in the ranking, as they will be swiftly reprimanded for attempting to play with a high-ranking group member (Burghardt, 2014). These corrections are vital to animals who must learn their place early on or face expulsion from the group.

As mentioned earlier, for species living in resource-rich environments, there will be moments of "down time." When play fills that time, it may improve an animal's behavioral and cognitive flexibility. It seems that engaging in large amounts of play—both socially and alone—is associated with improved adaptation to changing environments (Kuczaj, Makecha, Trone, Paulos, & Ramos, 2006; Kuczaj & Walker, 2006). This might be because play requires the individual to respond to the actions of an unpredictable object or social partner. When a wrestling partner zigs instead of predictably zagging or a toy squeaks out of nowhere, the animal's mind starts to work—there's no evolutionary script for an unexpected zig or a squeaking toy. The neurons in the brain grow and add connections every time something is learned, so through play, the animal can integrate new, lasting understandings about their environment. Among primates, for example, frequency of play is directly related to both brain size and behavioral flexibility, further illustrating the cognitive benefits of play (Deaner, Isler, Burkart, & van Schaik, 2007; Iwaniuk et al., 2001; Montgomery, 2014).

Because play is cognitively stimulating, humans who care for animals often try to design ways to promote play. In the wild, cognitive resources are focused on basic needs—defending territory, finding food, avoiding predators, and so on. These needs are met by us when animals live in our care, so their brains must be stimulated in other ways. All zoos and aquariums accredited by the Association of Zoos and Aquariums—and many that are not large enough for accreditation—are required to provide enrichment and opportunities for play to the animals in their care (*Accreditation Standards and Related Policies*, 2017). As program directors know, engaging in play behavior is considered a strong indicator of good health and welfare (Held & Špinka, 2011). Creating intentional opportunities for animals to interact with toys and humans can enhance the cognitive well-being of animals in human care by encouraging spontaneous behaviors, problem solving, and simulation of wild play behavior (Shepardson, Mellen, & Hutchins, 1998). With that said, if you have animals at home, now is a good time to pause and go play with them. As research has shown, you'll both be better off for doing so!

INNOVATION

It's an early morning in London in 1949, and you stretch, yawn, and step outside to grab the day's milk. Much to your surprise, someone has beaten you to it, and it's not that your partner has woken up early and brought in the milk for you. No, there on your front stoop are your milk bottles, open and slightly drained, with suspicious drops of milk dotting the porch.

Believe it or not, you have been the victim of the first recorded instance of transmission of a novel behavior by an animal population (Fisher & Hinde, 1949). At some point in the past few days or week, a small bird learned how to open those milk bottles and was rewarded with a wonderful treat. Soon, other birds came to watch. They, too, learned to open the milk bottles. And so on. Fifteen years later in Japan, another extraordinary case of animal innovation is documented—that of Imo the Japanese macaque. Not only did Imo's life hack of rinsing sand off her sweet potatoes before eating them spread among her conspecifics, but her future creative endeavors also drew a following. For example, when scientists scattered rice into the sand on the beach, Imo developed a smart solution by tossing handfuls into the water and skimming the floating rice from the water's surface (Kawai, 1965).

Since these two cases, several other species have become the focus of animal innovation research. Human researchers define creativity as something that is both novel and appropriate to the task (Kaufman, 2016). You might do something extremely unusual—build a four-story house out of crackers, for example—but unless it's appropriate to the task (of building a house), it's not creative, it's chaos. In the field of animal cognition, appropriateness is generally measured by survival—something that is novel but not appropriate is much more likely to get you killed—more on that later.

As you might expect, studying innovation in animals is even more challenging than it is in humans (which is already a difficult task; see Kaufman, 2016). The creative process is awkward to observe in experimental conditions in humans ("Be creative! Go! Right now!"), and with the exception of the next example, it becomes more problematic when a communication barrier is added (Kaufman & Kaufman, 2014). At Sea Life Park in Hawaii, animal trainer Karen Pryor tested her prediction that dolphins could not only innovate but plan to innovate and do so in a "Be creative! Go!" kind of way. To test this, she actually trained a group of dolphins to do behaviors that they had not previously done when given a particular command (Pryor, Haag, & O'Reilly, 1969). In doing so, Pryor and her staff pioneered the "do something new" command, allowing for creativity sessions in which animals offered behaviors and were reinforced for being creative. By requesting "do something new" several times in succession, the staff at Sea Life Park was able to elicit increasingly complex behaviors, such as multiple back or front flips and synchronized behaviors when two animals were simultaneously presented with the command. This type of training for innovative behaviors is still used in many training programs today in zoos, aquariums, and with pets (Pryor, 2014), and it is a particularly good way to stimulate cognitive abilities in animals as well as being fun for humans to see what their animals create. Further, it offers researchers a way around the communication barrier by giving the animal a command to be creative on cue.

Most research on animal creativity is not as convenient as Pryor's dolphin work, so we do not have much choice but to either wait for innovation to naturally occur or set up situations to try to elicit innovation. Then, we apply human standards to evaluate whether innovation has actually occurred. Very often this means we can only study what is called the creative product—the tool or behavior that is created and, in many cases, may not last longer than the time during which it is in use. As a result, several subtopics tend to serve as surrogates for creativity in animals—innovation frequency, social learning, behavioral transmission, neophilia (i.e., seeking out novel or new objects or experiences), play, and personality are closely tied to creative abilities yet much easier to examine empirically (Kaufman & Kaufman, 2014). At the species level, researchers might use the number of times a particular species has been observed creating tools as a way to gauge which species of bird is more creative than the other (Lefebvre, Whittle, Lascaris, & Finklestein, 1997), while a laboratory experiment might measure how fearful rats are of novel objects as way to measure individual differences in creative predispositions (Dellu, Piazza, Mayo, Le Moal, & Simon, 1996).

Being able to distinguish novel items and scenarios as well as having a lack of fear of them seem to be foundational dispositional qualities for innovation. Neophilia and risk taking can be viewed as similar to the human personality trait "openness to experience," and attraction to novel objects and situations is positively correlated with innovation in a wide variety of species, from birds to marine mammals (Day, Coe, Kendal, & Laland, 2003; Krueger, Farmer, & Heinze, 2014; Kuczaj et al., 2009). However, animals must strike a delicate balance between pursuing new things and sticking with the tried-and-true conservative ones that are characteristic of their species (Kaufman & Kaufman, 2014). Group-living animals, species of ants and birds, for example, have been shown to thrive with very specific ratios of risk-takers to non-risk-takers in the group—enough risk-takers to find new food sources, but not enough to gamble the well-being of the group or colony (Heck & Ghosh, 2000, 2002).

In the laboratory, recognition of novelty, and fear or attraction to it, is most often studied by the introduction of an animal to a new food, situation, or object. In the open-field test, an animal is placed in a new, open environment, and behaviors like the number and location of "rearing" behaviors are tracked (Belzung, 2010). Rearing in this context, which involves standing on the hind feet, allows animals to get a better look at the immediate area. It is, however, is potentially dangerous due to exposure to predators. Researchers can also study the relationship between genes and the environment by engineering strains of high- and low-novelty rats (e.g., Dellu, Mayo, Piazza, Le Moal, & Simon, 1993). The high-novelty rats are more likely to engage in risky behaviors like rearing, are more attracted to novelty, and tend toward sensation seeking. These rats are also more likely to investigate novel areas and more of them—perhaps for reasons akin to human boredom or curiosity. The parallels to human sensation seeking, up to and including the intrinsically reinforcing qualities of the so-called adrenaline rush, are impossible to ignore, which makes them an interesting model for studying the biology underlying creative dispositions.

Since we cannot ask animals to self-report on any intrinsic reward they experience when they explore novel objects or innovate, the best we can do is speculate on creativity's benefits to the species' survival. Making tools that improve foraging for food is one place where animals often innovate. For example, tuskfish (*Choerodon schoenleinii*) have been observed using rocks as a surface on which to break open cockles (Jones, Brown, & Gardner, 2011). Another place to look for survival-based innovation is in the behavior of invasive animal species. An invasive species is one that is not native to a particular geographic area but is able to be successful. An excellent example of an invasive animal species is the Norway, or brown, rat, which arrived in the United States with British colonists in the late 1700s. Chances are, if you live in the United States and you've seen a rat scamper across an alley at night, or scamper through a subway tunnel, or scamper through your basement, or really scamper anywhere, you saw a Norway rat. Like any invasive species, in order to succeed in their new home, Norway rats had to be willing to test out new food sources and quickly learn what was safe to eat, secure proper shelter using new materials, and develop strategies for evading new types of predator (Audet & Lefebvre, 2017; Beever et al., 2017). To say they have accomplished all of these is a real understatement.

The Australian bowerbird (*Ptilonorhynchidae* spp.) is a well-known final animal example of how creativity is marked by both novelty and appropriateness. Males create elaborate, unique nests, called bowers, which are decorated with items the bird has scavenged for—the shinier, the better—and some species even have specific color preferences (Doerr, 2012). Once the bower is completed, the male stands outside his bower and dances to attract a female. In bowerbirds, females choose who to mate with by judging the best bower and dance (Endler & Day, 2006). Shiny bowers and creative dances earn more females' attention, but an overly creative courtship dance might result in a display that is more frightening to a female than it is impressive, thus scaring away a potential mate (Patricelli, Coleman, & Borgia, 2006). Humans are usually willing to forgive a bad dancer, but for animals like the bowerbird, too creative of a dance is damning to a male's reproductive success (Kaufman, 2016). This is where the stipulation of "appropriateness" becomes significant. While creativity and behavioral flexibility can drastically improve an animal's situation, inappropriate creativity can result in dying without passing on one's genes. Innovation in the animal world is a far more delicate balance, and there are rarely second chances.

ANIMAL SPOTLIGHT: BETTY THE CROW

In 2002, a New Caledonian crow (*Corvus moneduloides*) by the name of Betty was one of the subjects in a laboratory tool use experiment, which involved deciding between using a straight wire or a hooked wire to retrieve a piece of food from a tube contraption. Another subject named Abel had run off with the hooked tool, so Betty took a look at the contraption, selected the straight piece of wire, and bent it into a hook of her own (Weir, Chappell, & Kacelnik, 2002). In subsequent testing, she repeated the behavior many times in order to obtain food. Amazingly enough, Betty had had no prior experience with tool making, hooks, or bending wires, and no models to watch (Weir et al., 2002). She essentially saw a problem and not only solved it but manipulated the structure of an object to make a customized tool that helped her solve it. This represents a step above Köhler's chimpanzees climbing on top of stacked crates or lizards overriding their pounce response by flipping a cap over. Even more, subsequent studies showed that Betty's tool use abilities improved over time and that she would modify her tools such as unbending as well as rebending wires (Weir & Kacelnik, 2006), illustrating a truly cognitive mind.

To be fair, Betty's not the first to create tools. Wild New Caledonian crows regularly make and use tools in their native New Caledonia, an archipelago off the coast of Australia. The bending behavior Betty used in the laboratory is carried out by her wild counterparts using twigs, which caused researchers to speculate that Betty's innovative tool use was at least in some part due to a biological predisposition to be attracted to objects that could be molded and used to extract food from tree logs or other natural substrates (Rutz, Sugasawa, van der Wal, Klump, & St Clair, 2016). However, a related corvid species, the rook, does not appear to use tools in the wild, but it does seem to share the talent of making and using tools in the laboratory, demonstrating how living in human care can change an animal's behavior. These birds will use stones to collapse an out-of-reach platform containing a worm, choose stones of the correct size and orientation, and transport the appropriate stones from distant locations (Bird & Emery, 2009).

It's not too surprising that a corvid would have done all this. There's a good chance you have a smart crow story of your own. I knew some researchers once who were working to radio track crows, but they could never catch the crows to fit them with transmitters. One of the common bird-capturing methods they used was to drop a net down onto them from above. Once the net comes down, most birds thrash around and get their beaks caught, at which point scientists go in and carefully untangle them. Anecdotally, my colleagues said the crows learned to "duck their heads down" to protect their beaks from snagging and run out from under the nets.

It's hard to deny that the majority of published studies on bird brain power focus on corvids and psittacines (parrots). Relative to chimpanzees, these birds seem to share many of the same cognitive abilities stemming from a common mental "toolbox" (Emery & Clayton, 2004). The behaviors you see in backyard birds like blue jays and crows are actually quite complex. Food caching, for example, requires an understanding of Piagetian object permanence—meaning the bird must understand that food still exists despite being out of sight (Hoffmann, Rüttler, & Nieder, 2011). It can also require an understanding of mental time travel—that a past, present, and future exist separately. Many animals cache food, but there is evidence that corvids are actually planning for the future when they do so, whereas in other cases, like winter acorn hoarding, it's more instinctual. For example, one study provided food to blue jays every morning in one of two compartments. While one compartment was filled, the second compartment was always left empty. After this routine was established, the birds were presented with both the compartment that was regularly filled and a second food in the previously empty compartment. They were allowed to eat and store food. Here, the jays showed a strong preference for caching this "new" food source in the container that had always gone empty, thus seeming to anticipate the regular filling of the other (Raby et al., 2007).

Mental time travel may have been directly involved in Betty's ability to plan the type of tool she needed to accomplish her task. Further reinforcing the point, Betty was the first animal on record to spontaneously use three tools in sequence to reach a goal. Food had been placed in an elongated tube, only accessible by a long stick. However, the stick of appropriate length had to be obtained by using a different stick (of a shorter length). In order to access the tool she needed to obtain the food, she had to plan for the future and anticipate what each of the tools would allow her to do. Betty was the only subject in the group to successfully complete the task, although several of the other crows in the study made at least some progress. It's important to note that the authors found evidence for individual differences in performance based on (1) how much experience each animal had with tools and tool use and (2) how engaged they were by the task itself, illustrating the dynamic relationship between genes and the environment when it comes to cognitive flexibility (Wimpenny, Weir, Clayton, Rutz, & Kacelnik, 2009).

When Rutz et al. (2016) documented innovative tool manufacturing and use by wild New Caledonian crows, it forced the examination of Betty's accomplishments in a new light. If stick bending is a natural behavior, then what Betty did was to apply something she may have instinctually been predisposed to do in a situation where it was not instinctual (there are not many treats in the bottom of tubes in the wild). Still, this does not negate what we learned—Betty still showed us that she could use this natural behavior in a novel setting, that she could evaluate her success and modify her approach accordingly, and that she appeared to be able to do this with an understanding of the present versus future conditions. It is also, however, a cautionary tale of why the focus of laboratory research should be moved out into the field and vice versa. When psychologists and biologists collaborate, a more informed understanding of animals' cognitive abilities emerges.

HUMAN APPLICATION: CREATIVITY IN EVERYDAY LIFE

Human creativity researchers call something creative if it is both novel and appropriate to the task (Sternberg, 1985). For this reason, creativity extends beyond simple randomness. To fully examine creativity, researchers explore what they call the "four Ps"—Press, Product, Process, and Person. We'll unpack each in turn, using examples and insights from the study of human creativity. Along the way, we'll also explore what science says about how to maximize creativity in your own life.

The first P, the creative Press, pertains to the surroundings in which creativity occurs. It examines whether creativity is fostered or encouraged, what resources are present (e.g., art supplies), and if there is an audience available to evaluate the finished product. Much of the research surrounding the creative Press occurs in educational settings. For example, what can we do to help children be more creative? Surprisingly, two of the biggest "creativity killers" in and out of the classroom are extrinsic motivation (i.e., rewards) for being creative and the anticipation of having the creative product be evaluated by someone else. Creative productivity is negatively correlated with extrinsic motivation, meaning the more motivated someone is by external rewards, the less creative their output tends to be. The lackluster album from a musician who is chasing fame and fortune, rather than writing songs because they love music, illustrates this relationship nicely. Likewise, a lack of outside expectation, judgment, or constraint improves creativity (Kaufman, 2016). Sound familiar? Or not, as the case may be? Very often the things our educational system encourages—extrinsic motivation, outside judgment, problem solving within a set of rules or guidelines—are exactly the wrong things to foster creativity for children and later for adults (Hennessey & Amabile, 1998).

The creative Product is what we most often use to study or judge creativity, perhaps because it's usually tangible. The creative product is simply what is produced by the creative act. Sometimes it's output we can easily point to, like "art"—a painting, song, or dance. However, a new technique to solve a mathematical proof or a novel combination of chemicals to make a new kind of plastic is also a creative product. The solution to a novel problem or predicament in life could also be a creative product, although it's certainly not tangible. When it comes to measuring a creative product, one method is called the consensual assessment technique (CAT), which involves at least three judges rating the product on scales like effort, neatness, novelty, and appropriateness (Amabile, Berglas, & Handel, 1982; Baer, Kaufman, & Gentile, 2004; Kaufman, Baer, Cole, & Sexton, 2008). Given the inherent nebulousness of the word *creativity*, you might expect the judges would be all over the place in their scores. On the contrary, when the judges are experts in the area they're evaluating (which the CAT requires), they tend to agree very closely with each other on how creative a product is—even if they cannot necessarily put into words why they think it's creative or even describe what they're looking for in the creative product to consider it so.

The creative Process is the hardest of the four to measure logistically (think "Be creative. 1-2-3, Go!"). The steps to achieve an innovative solution to a conflict, a new spin on an old recipe, a novel work-around, and creative ways of looking at the world are not easily recorded—but they're all things we're familiar with. Part of the reason why it's hard to introspect on the creative process in the moment is because when you're totally immersed in the creative process, you may find yourself in what creativity researchers describe as creative flow. This state is characterized by losing track of time and surroundings; everything melts away, and you're in the zone. Measuring the process of creativity is like asking someone to be in the zone, but not so far in the zone that they cannot report on their experience of being in the zone, and therein lies the challenge (Csikszentmihalyi, 1996).

Finally, the creative Person is the one who puts it all together to actually produce the creative act. Things start to get into more technicalities and specifics when we want to define or learn about the creative person, which can be further divided into four Cs—Big-C, Pro-c, little-c, and mini-c. When asked to name someone who's creative, most people tend to go for famous people in the arts like Wolfgang Mozart, Frida Kahlo, Emily Dickinson, or Michael Jackson. Though not always traditional artists, Big-C creators are those who are highly prolific, legendary in their field, remembered by the general public, and usually deceased (Kaufman & Beghetto, 2009). They are often studied via historiometric methods, which involve examining biographies and diaries to track their lives and creative accomplishments (Simonton, 2009).

You and I will probably never be Big-C creators. And that's okay—in fact, it could be harmful to compare ourselves to Big-C creators. It's fairly easy to fall into the trap of "If I can't be Mozart, I might as well not learn to play piano." While you may not reach Big-C status, you could become very well known within your community for something you create. You mention a new recipe to a friend, who passes it on to another friend, and suddenly people are asking you when the restaurant is going to open. Pro-c, or professional level creativity, comes when you are known in a field for something. You may not be world famous, but among other chefs you have an excellent reputation (Kaufman & Beghetto, 2009). Some of the individual animals discussed in the chapters of this book are the equivalent of Pro-c—Alex the parrot, Phoenix and Akeakami the dolphins, Kanzi the bonobo, and Betty the crow (Kaufman & Kaufman, 2014). Among animal cognition researchers, these are "rock stars"—anyone who studies animal cognition knows the names and their accomplishments. It would not be professional for us to admit to the likelihood of passing out if given the opportunity to meet Kanzi, but...

For most of us, the non-Mozarts and non-Kanzis of the world, you and I are engaged in everyday creativity—the type of creativity we encounter all the time just going about our lives. This kind of creativity is referred to as little-c creativity. When you take an art class, create a beautiful floral arrangement, or make up a song to entertain your toddler, you are a little-c creator (Kaufman & Beghetto, 2009). The first chimpanzee to ever use a stick to fish for termites and the very first dolphin to carry a piece of sponge on her nose when foraging were also little-c creators. They (and you) were using creativity to enhance everyday life (Kaufman & Kaufman, 2004). Not everyone will love your floral arrangement, and the stick might not work on every termite mound, but it works for you in that moment, and chances are it, or a variation of it, will work for at least a couple other individuals as well, thereby enhancing their lives.

Lastly, creativity can be personally meaningful to you and you only. This is mini-c creativity (Kaufman & Beghetto, 2009). The poetry you wrote in your diary in eighth grade was creative, insightful, and extremely meaningful. To you. Emphasis on the *you*. My husband once did a study on mini-c, which required teachers to read hundreds of poems written by eighth graders. Those teachers do not return his calls anymore. Mini-c creativity does not have to be shared with anyone; it can be for you and you alone. It can also promote positive well-being. While there are connections between creativity and mental illness, it's impossible to tell if being creative makes a person more susceptible to mental illness or if people with mental illness are more creative (Schlesinger, 2009). While we may never know the answer to that question, there is evidence for what has been dubbed "the Writing Cure," where certain types of creative output, specifically narrative writing, can actually decrease a person's susceptibility to depression (Pennebaker, 1997).

As humans, we're lucky that we do not always have to be focused on our own life or death survival. We can use that extra energy and cognitive resources to enrich our own lives with everyday creative acts. But how can we do that? First off, remember a few things. To begin, creativity is not just "art"; you can be creative in anything (Kaufman & Baer, 2005). Cooking, doodling, creating nature walks, developing cures for diseases. Always remember little-c (Kaufman & Beghetto, 2009). You do not have to be Picasso, Julia Child, or Ada Lovelace.[4] Remember that the best creativity is creativity that's done *for yourself*—intrinsically motivated creative output (Baer, 1998; Hennessey & Amabile, 1998). Create something that makes you happy, makes you feel good, or makes you proud. But do not forget to do your homework. "Inspiration" is overrated. You have to know something about a subject to be creative in it (Cropley, 2006). No one ever became a creative chef without learning how brown sugar makes a recipe taste different from white sugar. Lastly, remember to try new things. Being open to experience, a trait that psychologists commonly use to assess personality, both intellectually and experientially is associated with higher creativity. The more variety you experience in life, the more variety you can produce (McCrae, 1987).

It's important to emphasize again that creativity is novel but also appropriate. Sometimes, it's not the right time to be creative. The ability to know when it's appropriate to be creative is called *creative metacognition*. And since creativity is by definition appropriate to the situation, a big part of being creative is knowing when to (and when not to) be creative (Kaufman & Beghetto, 2013). But in the end, it all pays off. Creative people are more likely to be happy with their

---

4. Considered the first computer programmer. Look her up; she's awesome.

lives, to get promotions, to be healthier, to be more resilient when problems arise—even to have higher salaries (Kaufman, 2016)! They love what they do, and it shows.

YOUR TURN!

Materials for animal cognition research can come from a variety of places, including your junk drawer! In this DIY research idea, we outline some methods for how you can test variations on the famous string-pull task in your own home with an animal subject of your choosing. Who knows, maybe your parrot will show herself to be a particularly insightful mini-c creative problem solver!

In order to conduct this study, you'll need some string, rope, or cord that is thick enough for the animal you'll be working with to grasp and manipulate it with teeth, paws, or beak. You'll also need a food-motivated animal, a device to video record the animal's behavior, and some treats. In the spirit of this chapter's "Human Application," some creative thinking will probably also be necessary along the way. As we described in more detail earlier in the chapter, the general task set up is to attach food to one end of a string so that it's out of the animal's reach. To solve the task, your subject will need to figure out a way to reel in the string.

Though the task is most commonly used with birds, we'll explain how you can modify it for a variety of different animals. Regardless of the species you work with, you'll want to use two strings. One of the strings, the experimental string, will have the food item tied to the end of it. The control condition string should have nothing tied to it. The purpose of this second empty string is to see whether the animal understands the relationship between the string it's interacting with and what's at the bottom of it. If the animal interacts with only the baited string, it strengthens the idea that its behavior was goal oriented toward solving the problem of reaching the food. If it interacts with both strings at random, it will suggest the animal does not understand the connection between the problem, its behavior, and the food getting closer.

If you're working with a bird, you can set up the classic string pull task using a perch with the two strings tied to it hanging down about 30 to 60 cm (1–2 ft), depending upon the size of the bird.

The two strings should be placed far apart enough to avoid them getting tangled, so about 15 cm (6 in.) or so.

If you're working with a terrestrial mammal, reptile, or amphibian, we encourage you to work horizontally rather than vertically for safety reasons. An upside-down laundry basket with the two strings fed through the holes could do the trick of protecting the food from being eaten while allowing your research subject to see the food is inside. Leave the ends of the strings outside of the basket within reach, and maybe set some books on top of the basket just in case your subject decides to try to flip it over.

If you're working with a fish, you are chartering new territory, since it does not seem like anyone has published details on how to do the task underwater. This is where the ingenuity comes in! Read the basic setup for the preceding species, and use it as inspiration for the fish. If you come up with something that works, let us know. We'd love to hear from you.

With your methodology worked out, you can bring your research subject into the testing area and start video recording. In this case, video recording is easier than live recording because as we've said before, behavior happens fast. Trying to keep up with the steps using hand-written notes would be exhausting. Some studies allow the animal to work on the problem until a solution is reached; others place a time limit like 10 minutes on each testing session. If more than 10 minutes go by, you might consider stopping. No one likes to be teased or frustrated. Remove the apparatus, and take a break. Redirect any frustration by playing together, going for a walk, giving some food rewards for other behaviors, and so on.

When reviewing your video footage, take note of the animal's general behaviors such as how active they are as time goes on, whether or not you see behaviors that might signal stress, what kind and how many vocalizations the animal emits, as well as any attempts to interact with or solicit help from you (e.g., eye gaze toward you and then to the problem to be solved, pawing at you). As a side note, while you're recording, if your animal tries to interact with you, ignore them as much as possible—which can be difficult, we know. Pay special attention to specific behaviors your subject makes to the strings. Does she bite or pull a string, stop, and move to get a better look or a different angle of the problem? Do you notice any other behaviors that signal "testing" the relationship between their behavior and the string like this? Do you notice any insightful problem solving, whereby the animal circles or observes the problem for a while and then goes straight to the correct string and retrieves the food? All of these observations are data—data that would be really hard to accurately represent without using this video coding method of analyzing video footage.

Continue with additional string-pulling trials if you'd like, following your same methods or modifying and trying again. For horizontal string tests, for example, you might replicate some previous studies that either cross one string over another or include a control string that has been cut in one or multiple places to see how the animal responds to these new challenges. As always, get creative, and remember to have fun—yourself and your research subject included!

REFERENCES

Accreditation Standards and Related Policies. (2017). Retrieved from https://www.aza.org/assets/2332/aza-accreditation-standards.pdf

Alem, S., Perry, C. J., Zhu, X., Loukola, O. J., Ingraham, T., Søvik, E., & Chittka, L. (2016). Associative mechanisms allow for social learning and cultural transmission of string pulling in an insect. *PLoS Biology, 14*(10), 1–28. doi:10.1371/journal.pbio.1002564

Amabile, T. M., Berglas, S., & Handel, M. (1982). *Social psychology of creativity: A consensual assessment technique.* Retrieved from https://product.design.umn.edu/courses/pdes2701/documents/5701papers/01creativity/amabile82.pdf

Audet, J., & Lefebvre, L. (2017). What's flexible in behavioral flexibility? *Behavioral Ecology, 28*(July), 943–947. doi:10.1093/beheco/arx007

Auersperg, A. M. I., van Horik, J. O., Bugnyar, T., Kacelnik, A., Emery, N. J., & von Bayern, A. M. P. (2015). Combinatory actions during object play in parrots (*Psittacus erithacus*) and corvids (*Corvus*). *Journal of Comparative Psychology, 1*, 62–71. doi:10.1037/a0038314

Azevedo, A. F., & Van Sluys, M. (2005). Whistles of tucuxi dolphins (*Sotalia fluviatilis*) in Brazil: Comparisons among populations. *Journal of the Acoustical Society of America, 117*(3), 1456–1464. doi:10.1121/1.1859232

Baer, J. (1998). Gender differences in the effects of extrinsic motivation on creativity. *Journal of Creative Behavior, 32*, 18–37. doi:10.1002/j.2162-6057.1998.tb00804.x

Baer, J., Kaufman, J. C., & Gentile, C. A. (2004). Extension of the consensual assessment technique to nonparallel creative products. *Creativity Research Journal, 16*(1), 113–117. doi:10.1207/s15326934crj1601_11

Bauer, E. B., & Smuts, B. B. (2007). Cooperation and competition during dyadic play in domestic dogs, Canis familiaris. *Animal Behaviour, 73*(3),

489–499. Retrieved from http://www.sciencedirect.com/science/article/B6W9W-4MR86MN-2/2/7ad0516747630b6ae9b9a8e0fcbe5f6f

Beever, E. A., Hall, L. E., Varner, J., Loosen, A. E., Dunham, J. B., Gahl, M. K., … Lawler, J. J. (2017). Behavioral flexibility as a mechanism for coping with climate change. *Frontiers in Ecology and the Environment.* doi:10.1002/fee.1502

Belzung, C. (2010). Open-field test. In I. P. Stolerman (Ed.), *Encyclopedia of Psychopharmacology* (pp. 920–926). New York, NY: Springer.

Benson-Amram, S., Weldele, M. L., & Holekamp, K. E. (2013). A comparison of innovative problem-solving abilities between wild and captive spotted hyaenas, Crocuta crocuta. *Animal Behaviour, 85,* 349–356. doi:10.1016/j.anbehav.2012.11.003

Bird, C. D., & Emery, N. J. (2009). Insightful problem solving and creative tool modification by captive nontool-using rooks. *PNAS, 106*(25), 10370–10375. Retrieved from http://www.pnas.org/content/106/25/10370.short

Bond, A. B., Diamond, J., & Bond, A. B. (2003). A comparative analysis of social play in birds. *Behaviour, 140*(8), 1091–1115. doi:10.1163/156853903322589650

Bond, A. B., Kamil, A. C., & Balda, R. P. (2007). Serial reversal learning and the evolution of behavioral flexibility in three species of North American corvids (Gymnorhinus cyanocephalus, Nucifraga columbiana, Aphelocoma californica). *Journal of Comparative Psychology, 121,* 372–379. doi:10.1037/0735-7036.121.4.372

Breland, K., & Breland, M. (1961). The misbehavior of organisms. *American Psychologist, 16*(11), 681–684. doi:10.1037/h0040090

Burghardt, G. M. (2005). *The genesis of animal play: Testing the limits.* Cambridge, MA: MIT Press.

Burghardt, G. M. (2014). A brief glimpse at the long evolutionary history of play. *Animal Behavior and Cognition, 1*(2), 90–98. doi:10.12966/abc.05.01.2014

Burghardt, G. M. (2015). Play in fishes, frogs and reptiles. *Current Biology, 25*(1), R9–R10. doi:10.1016/j.cub.2014.10.027

Burghardt, G. M., Ward, B., & Rosscoe, R. (1996). Problem of reptile play: Environmental enrichment and play behavior in a captive Nile soft-shelled turtle,Trionyx triunguis. *Zoo Biology, 15*(3), 223–238. doi:10.1002/(SICI)1098-2361(1996)15:3<223::AID-ZOO3>3.0.CO;2-D

Caro, T. M. (1995). Short-term costs and correlates of play in cheetahs. *Animal Behaviour, 49,* 333–345. doi:10.1006/anbe.1995.9999

Ciani, F., Dall'Olio, S., Stanyon, R., & Palagi, E. (2012). Social tolerance and adult play in macaque societies: A comparison with different human cultures. *Animal Behaviour, 84*(6), 1313–1322. doi:10.1016/j.anbehav.2012.09.002

Clayton, N. S., & Dickinson, A. (1998). Episodic-like memory during cache recovery by scrub jays. *Nature, 395,* 272–274. doi:10.1038/26216

Colbert-White, E. N., Covington, M. A., & Fragaszy, D. M. (2011). Social context influences the vocalizations of a home-raised African grey parrot (*Psittacus erithacus erithacus*). *Journal of Comparative Psychology, 125*(2), 175–184. doi:10.1037/a0022097

Colbert-White, E. N., McCord, E. M., Sharpe, D. L., & Fragaszy, D. M. (2013). String-pulling behavior in the Harris's Hawk. *Ibis, 155,* 611–615. doi:10.1111/ibi.12040

Correia, S. P. C., Dickinson, A., & Clayton, N. S. (2007). Western scrub-jays anticipate future needs independently of their current motivational state. *Current Biology, 17*(10), 856–861. doi:10.1016/j.cub.2007.03.063

Cropley, A. (2006). In praise of convergent thinking. *Creativity Research Journal, 18*(3), 391–404. doi:10.1207/s15326934crj1803_13

Crystal, J. D. (2013). Remembering the past and planning for the future in rats. *Behavioural Processes, 93,* 39–49. doi:10.1016/j.beproc.2012.11.01

Csikszentmihalyi, M. (1996). *Creativity: Flow and the psychology of discovery and invention.* New York, NY: Harper Collins.

Day, R. L., Coe, R. L., Kendal, J., & Laland, K. N. (2003). Neophilia, innovation and social learning: A study of intergeneric differences in callitrichid monkeys. *Animal Behavior, 65*(3), 559–572. doi:10.1006/anbe.2003.2074

De la Fuente, I. M., Bringas, C., Malaina, I., Fedetz, M., Perez-Samartin, A., Lopez, J. I., ... & Boyano, M. D. (2018). Evidences of conditioned behavior in Amoeba proteus. *bioRxiv,* 264176. doi:10.1101/264176

Deaner, R. O., Isler, K., Burkart, J. M., & van Schaik, C. P. (2007). Overall brain size, and not encephalization quotient, best predicts cognitive ability across non-human primates. *Brain, Behavior and Evolution, 70*(2), 115–124. Retrieved from http://www.karger.com/DOI/10.1159/000102973

Dellu, F., Mayo, W., Piazza, P. V, Le Moal, M., & Simon, H. (1993). Individual differences in behavioral responses to novelty in rats. Possible relationship with the sensation-seeking trait in man. *Personality and Individual Differences, 15*(4), 411–418. doi:10.1016/0191-8869(93)90069-F

Dellu, F., Piazza, P. V, Mayo, W., Le Moal, M., & Simon, H. (1996). Novelty-seeking in rats--Biobehavioral characteristics and possible relationship with the sensation-seeking trait in man. *Neuropsychobiology, 34*(4), 136–145. doi:10.1159/000119305

Doerr, N. R. (2012). Male great bowerbirds accumulate decorations to reduce the annual costs of signal production. *Animal Behaviour, 83*(6), 1477–1482. doi:10.1016/j.anbehav.2012.03.021

Emery, N. J., & Clayton, N. S. (2004). The mentality of crows: Convergent evolution of intelligence in corvids and apes. *Science (New York, N.Y.), 306*(5703), 1903–1907. doi:10.1126/science.1098410

Endler, J. A., & Day, L. B. (2006). Ornament colour selection, visual contrast and the shape of colour preference functions in great bowerbirds, Chlamydera nuchalis. *Animal Behaviour, 72*(6), 1405–1416. Retrieved from http://www.sciencedirect.com/science/article/B6W9W-4M4KR29-1/2/1b1b74b30d3aa3f60aecc55863bc2069

Epstein, R., Kirshnit, C., Lanza, R., & Rubin, L. (1984). 'Insight' in the pigeon: Antecedents and determinants of an intelligent performance. *Nature, 308*(1), 61–62. Retrieved from http://www.nature.com/nature/journal/v308/n5954/abs/308061a0.html

Finn, J. K., Tregenza, T., & Norman, M. D. (2009). Defensive tool use in a coconut-carrying octopus. *Current Biology : CB, 19*(23), R1069–R1070. doi:10.1016/j.cub.2009.10.052

Fisher, J., & Hinde, R. A. (1949). The opening of milk bottles by birds. *British Birds, 42*, 347–357. Retrieved from https://britishbirds.co.uk

Gick, M. L., & Holyoak, K. J. (1980). Analogical problem solving. *Cognitive Psychology, 12*(3), 306–355. doi:10.1016/0010-0285(80)90013-4

Hare, B. A., Brown, M., Williamson, C., & Tomasello, M. (2002). The domestication of social cognition in dogs. *Science, 298*(5598), 1634–1636. doi:10.1126/science.1072702

Harland, D. P., & Jackson, R. R. (2004). Portia perceptions: The *umwelt* of an araneophagic jumping spider. In F. R. Prete (Ed.), *Complex worlds from simpler nervous systems* (pp. 5–40). Cambridge, MA: MIT Press.

Heck, P. S., & Ghosh, S. (2000). A study of synthetic creativity: Behavior modeling and simulation of an ant colony. *IEEE Expert Intelligent Systems and Their Applications, 15*(6), 58–66. doi:10.1109/5254.895863

Heck, P. S., & Ghosh, S. (2002). The design and role of synthetic creative traits in artificial ant colonies. *Journal of Intelligent & Robotic Systems, 33*(4), 343–370. doi:10.1023/A:1015552602374

Heinrich, B. (1995). An experimental investigation of insight in Common Ravens (*Corvus corax*). *The Auk, 112*, 994–1003. doi:10.2307/4089030

Held, S. D. E., & Špinka, M. (2011). Animal play and animal welfare. *Animal Behaviour, 81*(5), 891–899. doi:10.1109/FTC.2016.7821622

Hennessey, B. A., & Amabile, T. M. (1998). Reality, intrinsic motivation, and creativity. *American Psychologist, 53*(6), 674–675. doi:10.1037/0003-066X.53.6.674

Hoffmann, A., Rüttler, V., & Nieder, A. (2011). Ontogeny of object permanence and object tracking in the carrion crow, Corvus corone. *Animal Behaviour, 82*(2), 359–367. Retrieved from http://www.sciencedirect.com/science/article/pii/S0003347211001977

Howard, A. M., & Fragaszy, D. M. (2014). Multi-step routes of capuchin monkeys in a laser pointer traveling salesman task. *American Journal of Primatology, 76*, 828–841. doi:10.1002/ajp.22271

Iwaniuk, A. N., Nelson, J. E., & Pellis, S. M. (2001). Do big-brained animals play more? Comparative analyses of play and relative brain size in mammals. *Journal of Comparative Psychology, 115*(1), 29–41. doi:10.1037/0735-7036.115.1.29

Japyassú, H. F., & Laland, K. N. (2017). Extended spider cognition. *Animal Cognition, 20*(3), 375–395. doi:10.1007/s10071-017-1069-7

Jelbert, S. A., Taylor, A. H., Cheke, L. G., Clayton, N. S., & Gray, R. D. (2014). Using the aesop's fable paradigm to investigate causal understanding of water displacement by new caledonian crows. *PLoS One, 9*(3). doi:10.1371/journal.pone.0092895

Jones, A. M., Brown, C. R., & Gardner, S. (2011). Tool use in the tuskfish Choerodon schoenleinii? *Coral Reefs*, (May), 4703–4703. doi:10.1007/s00338-011-0790-y

Kaufman, A. B., & Kaufman, J. C. (2014). Applying theoretical models on human creativity to animal studies. *Animal Behavior and Cognition, 1*(1), 78–90. doi:10.12966/abc.02.01.2014

Kaufman, J. C. (2016). *Creativity 101* (2nd ed.). New York, NY: Springer.

Kaufman, J. C., & Baer, J. (2005). *Creativity across domains: Faces of the muse.* Mahwah, NJ: Lawrence Erlbaum.

Kaufman, J. C., Baer, J., Cole, J. C., & Sexton, J. D. (2008). A comparison of expert and nonexpert raters using the consensual assessment technique. *Creativity Research Journal, 20*(2), 171–178. doi:10.1080/10400410802059929

Kaufman, J. C., & Beghetto, R. A. (2009). Beyond big and little: The Four C Model of Creativity. *Review of General Psychology, 13*, 1–12. doi:10.1037/a0013688

Kaufman, J. C., & Beghetto, R. A. (2013). In praise of Clark Kent: Creative metacognition and the importance of teaching kids when (not) to be creative. *Roeper Review, 35*, 155–165. doi:10.1080/02783193.2013.799413

Kaufman, J. C., & Kaufman, A. B. (2004). Applying a creativity framework to animal cognition. *New Ideas in Psychology, 22*(2), 143–155. doi:10.1016/j.newideapsych.2004.09.006

Kaufman, S. B., & Kaufman, J. C. (2007). Ten years to expertise, many more to greatness: An investigation of modern writers. *Journal of Creative Behavior, 41*(2), 114–124. doi:10.1002/j.2162-6057.2007.tb01284.x

Kawai, M. (1965). Newly-acquired pre-cultural behavior of the natural troop of Japanese monkeys on Koshima islet. *Primates, 6*(1), 1–30. doi:10.1007/BF01794457

Klump, B. C., Sugasawa, S., St Clair, J. J. H., & Rutz, C. (2015). Hook tool manufacture in New Caledonian crows: Behavioural variation and the influence of raw materials. *BMC Biology, 13*(1), 97. doi:10.1186/s12915-015-0204-7

Köhler, W. (1925). *The mentality of apes.* London, UK: Routledge & Kegan Paul Limited.

Krueger, K., Farmer, K., & Heinze, J. (2014). The effects of age, rank and neophobia on social learning in horses. *Animal Cognition, 17*(3), 645–655. doi:10.1007/s10071-013-0696-x

Kuczaj, S. A., & Eskelinen, H. C. (2014a). The "Creative Dolphin" revisited : What do dolphins do when asked to vary their behavior ? *Animal Behavior and Cognition, 1*(1), 66–77. doi:10.12966/abc.02.05.2014

Kuczaj, S. A., & Eskelinen, H. C. (2014b). Why do dolphins play? *Animal Behavior and Cognition, 1*(2), 113. doi:10.12966/abc.05.03.2014

Kuczaj, S. A., Gory, J. D., & Xitco Jr., M. J. (2009). How intelligent are dolphins? A partial answer based on their ability to plan their behavior when confronted with novel problems. *Japanese Journal of Animal Psychology, 59*(1), 99–115. doi:10.2502/janip.59.1.9

Kuczaj, S. A., Makecha, R., Trone, M., Paulos, R. D., & Ramos, J. A. (2006). Role of peers in cultural innovation and cultural transmission: Evidence from the play of dolphin calves. *International Journal of Comparative Psychology, 19*(2), 223–240. Retrieved from http://escholarship.org/uc/item/4pn1t50s.pdf

Kuczaj, I. I., Stan, A., Xitco Jr, M. J., & Gory, J. D. (2010). Can Dolphins Plan their Behavior?. *International Journal of Comparative Psychology, 23*, 664–670. Retrieved from https://escholarship.org/uc/uclapsych_ijcp

Kuczaj, S. A., & Walker, R. T. (2006). How do dolphins solve problems? In E. A. Wasserman & T. R. Zentall (Eds.), *Comparative cognition: Experimental exlorations of animal intelligence* (pp. 580–601). New York, NY: Oxford University Press.

Leal, M., & Powell, B. J. (2012). Behavioural flexibility and problem-solving in a tropical lizard. *Biology Letters, 8*, 28–30. doi:10.1098/rsbl.2011.0480

Lefebvre, L., Whittle, P., Lascaris, E., & Finklestein, A. (1997). Feeding innovations and forebrain size in birds. *Animal Behavior, 53*, 549–560. doi:10.1006/anbe.1996.0330

MacLean, E. L., Merritt, D. J., & Brannon, E. M. (2008). Social complexity predicts transitive reasoning in prosimian primates. *Animal Behaviour, 76*(2), 479–486. doi:10.1016/j.anbehav.2008.01.025

McCrae, R. R. (1987). Creativity, divergent thinking, and openness to experience. *Journal of Personality and Social Psychology, 52*, 1258–1265. doi:10.1037/0022-3514.52.6.1258

Menda, G., Shamble, P. S., Nitzany, E. I., Golden, J. R., & Hoy, R. R. (2014). Visual perception in the brain of a jumping spider. *Current Biology, 24*(21), 2580–2585. doi:10.1016/j.cub.2014.09.029

Mendes, N., Hanus, D., & Call, J. (2007). Raising the level: Orangutans use water as a tool. *Biology Letters, 3*(5), 453–455. doi:10.1098/rsbl.2007.0198

Montgomery, S. H. (2014). The relationship between play, brain growth and behavioural flexibility in primates. *Animal Behaviour, 90*, 281–286. doi:10.1016/j.anbehav.2014.02.004

Patricelli, G. L., Coleman, S. W., & Borgia, G. (2006). Male satin bowerbirds, Ptilonorhunchus violaceus, adjust their display intensity in response to female startling: An experiment with robotic females. *Animal Behavior, 71*(1), 49–59. doi:10.1016/j.anbehav.2005.03.029

Pennebaker, J. W. (1997). Writing about emotional experiences as a therapeutic process. *Psychological Science, 8*(3), 162–166. doi:10.1111/j.1467-9280.1997.tb00403.x

Pepperberg, I. M. (2004). "Insightful" string-pulling in Grey parrots (*Psittacus erithacus*) is affected by vocal competence. *Animal Cognition, 7*, 263–266. doi:10.1007/s10071-004-0218-y

Pryor, K. W. (2014). Historical perspectives: A dolphin journey. *Aquatic Mammals, 40*(1), 104–114. doi:10.1578/AM.40.1.2014.104

Pryor, K. W., Haag, R., & O'Reilly, J. (1969). The creative porpoise: Training for novel behavior. *Journal of the Experimental Analysis of Behavior, 12*(4), 653. Retrieved from http://www.ncbi.nlm.nih.gov/pmc/articles/PMC1338662/

Raby, C. R., Alexis, D. M., Dickinson, A., & Clayton, N. S. (2007). Planning for the future by western scrub-jays. *Nature, 445*(7130), 919–921. doi:10.1038/nature05575

Robinson, G. E., & Barron, A. B. (2017). Epigenetics and the evolution of instincts. *Science, 356*(6333), 26–27. doi:10.1126/science.aam6142

Rowley, I., & Chapman, G. (1986). Imprinting and learning in two sympatric species of cockatoo. *Behaviour, 96*, 1–16. doi:10.1163/156853986X00180

Rutz, C., Sugasawa, S., van der Wal, J. E. M., Klump, B. C., & St Clair, J. J. H. (2016). Tool bending in New Caledonian crows. *Royal Society Open Science, 3*(8), 160439. doi:10.1098/rsos.160439

Salwiczek, L. H., & Bshary, R. (2011). Cleaner wrasses keep track of the "when" and "what" in a foraging task. *Ethology, 117*(11), 939–948. doi:10.1111/j.1439-0310.2011.01959.x

Salwiczek, L. H., Prétôt, L., Demarta, L., Proctor, D., Essler, J., Pinto, A. I., ... Bshary, R. (2012). Adult cleaner wrasse outperform capuchin monkeys, chimpanzees and orang-utans in a complex foraging task derived from cleaner - Client Reef Fish Cooperation. *PLoS One, 7*(11), e49068. doi:10.1371/journal.pone.0049068

Schlesinger, J. (2009). Creative mythconceptions: A closer look at the evidence for "mad genius" hypothesis. *Psychology of Aesthetics Creativity and the Arts, 3*, 62–72. doi:10.1037/a0013975

Schuster, S., Rossel, S., Schmidtmann, A., Jäger, I., & Poralla, J. (2004). Archer fish learn to compensate for complex optical distortions to determine the absolute size of their aerial prey. *Current Biology, 14*(17), 1565–1568. doi:10.1016/j.cub.2004.08.050

Shepardson, J., Mellen, J. D., & Hutchins, M. (1998). *Second nature: Environmental enrichment for captive animals.* Washington, DC: Smithsonian.

Simonton, D. K. (2009). *Genius 101.* New York, NY: Springer Publishing Company.

Stanton, L., Davis, E., Johnson, S., Gilbert, A., & Benson-Amram, S. (2017). Adaptation of the Aesop's Fable paradigm for use with raccoons (*Procyon lotor*): Considerations for future application in non-avian and non-primate species. *Animal Cognition, 20*(6), 1147–1152. doi:10.1007/s10071-017-1129-z

Sternberg, R. J. (1985). Implicit theories of intelligence, creativity, and wisdom. *Journal of Personality and Social Psychology, 49*, 607–627. doi:10.1037/0022-3514.49.3.607

Tanaka, K. D., Morimoto, G., Stevens, M., & Ueda, K. (2011). Rethinking visual supernormal stimuli in cuckoos: Visual modeling of host and parasite signals. *Behavioral Ecology, 22*(5), 1012–1019. doi:10.1093/beheco/arr084

Tarsitano, M. S., & Jackson, R. R. (1994). Jumping spiders make predatory detours requiring movement away from prey. *Behaviour, 131*, 165–173. doi:10.1163/156853994X00217

Thorndike, E. L. (1898). Animal intelligence: An experimental study of the associative processes in animals. *The Psychological Review: Monograph Supplements, 2*, i-109. doi:10.1037/h0092987

Timberlake, W., & Lucas, G. A. (1989). Behavior systems and learning: From misbehavior to general principles. In S. B. Klein & R. R. Mowrer (Eds.), *Contemporary learning theories: Instrumental conditioning theory and the impact of biological constraints on learning* (pp. 237–275). Hillsdale, NJ: Lawrence Erlbaum Associates, Inc.

Trut, L. N. (1999). Early canid domestication: The farm-fox experiment. *American Scientist, 87*(2), 160–169.

Valero, A., & Byrne, R. W. (2007). Spider monkey ranging patterns in Mexican subtropical forest: Do travel routes reflect planning? *Animal Cognition, 10*, 305–315. doi:10.1007/s10071-006-0066-z

van Schaik, C. P., Damerius, L., & Isler, K. (2013). Wild orangutan males plan and communicate their travel direction one day in advance. *PLoS One, 8*, e74896. doi:10.1371/journal.pone.0074896

Waroff, A. J., Fanucchi, L., Robbins, C. T., & Nelson, O. L. (2017). Tool use, problem-solving, and the display of stereotypic behaviors in the brown bear (*Ursus arctos*). *Journal of Veterinary Behavior: Clinical Applications and Research, 17*, 62–68. doi:10.1016/j.jveb.2016.11.003

Weir, A. A. S., Chappell, J., & Kacelnik, A. (2002). Shaping of hooks in New Caledonian crows. *Science, 297*(5583), 981. doi:10.1126/science.1073433

Weir, A. A. S., & Kacelnik, A. A. (2006). A New Caledonian crow (*Corvus moneduloides*) creatively re-designs tools by bending or unbending aluminium strips. *Animal Cognition, V9*(4), 317–334. doi:10.1007/s10071-006-0052-5

Wilkinson, A., & Glass, E. (2018). 18 tortoises–cold-blooded cognition: How to get a tortoise out of its shell. *Field and Laboratory Methods in Animal Cognition: A Comparative Guide*, 401–419. doi:10.1017/9781108333191.020

Wimpenny, J. H., Weir, A. A. S., Clayton, L., Rutz, C., & Kacelnik, A. (2009). Cognitive processes associated with sequential tool use in New Caledonian crows. *PLoS One, 4*(8), e6471. doi:10.1371/journal.pone.0006471

Yi, X., Plucker, J. A., & Guo, J. (2015). Modeling influences on divergent thinking and artistic creativity. *Thinking Skills and Creativity, 16*, 62–68. doi:10.1016/j.tsc.2015.02.002

Zylinski, S. (2015). Fun and play in invertebrates. *Current Biology, 25*(1), R10–R12. doi:10.1016/j.cub.2014.09.068

Individual Differences Between Animals

My (AK) family once had a rescued iguana named Pancho.[1] Pancho had been kept in a glass tank in a kindergarten classroom without the proper lighting, so he'd never grown to his full size, and his health was not all that great either. He'd let me hold him, but not really anyone else. In fact, we were pretty sure Pancho hated everyone except me. If you asked anyone in my family, Pancho had personality and emotions. It's common for humans to ascribe personality and other individual differences to animals because that's how we relate to one another—how we get to know the diversity of individuals in our communities. While there's nothing fundamentally wrong with doing this with the animals in our lives, for most of the field's history, talking about individual differences in animals was avoided. To admit there were individual differences in animals' behavior and cognition would introduce a level of subjectivity into the science we

---

1. Actually, it turned out later that he was probably Panchette.

do. We've come a long way since then, and research has shown that individuals within a species are more different than we once believed—lending scientific credibility to what most of us already knew.

Our decision to include this chapter and where it's positioned in the book was intentional. Personality and emotional well-being (e.g., coping style) are features of an animal's behavior that are very real but have historically not been considered cognitive (Griffin, Tebbich, & Bugnyar, 2017). Intelligence, by nature of the word, implies cognition, but it is studied more by researchers via methods of general assessment rather than specific abilities like problem solving. Still, all three of the topics in this chapter are similar in that they act to shape the cognitive abilities we see. An animal that is biologically predisposed to have a shy personality may be less likely to innovate. Individual differences act as ever-present filters, influencing the cognitive processes that we have shared up to this point (Griffin et al., 2017). Including this chapter at the end of the book should serve to contextualize all of the findings. Most of the species participating in the studies we have described, even within a given experiment, have been shown to display individual differences of some kind. The more cognitively complex a species is, the more ways it can be affected—dispositionally, emotionally, intellectually—and, therefore, the more pronounced individual differences may seem.

As mentioned earlier, the study of individual differences has not always been a consideration in animal cognition research. This is because, like psychology, the goal is to develop principles that generalize across all individuals. By testing 150, 20, or even five individuals and combining their performance on cognitive tasks, researchers have historically tried to get an idea of how an "average" animal from a given species will perform. In doing so, we lose focus of individual differences (Shaw & Schmelz, 2017). For example, while it's a fascinating conclusion that magpies can pass the mirror-mark test, in some ways, I'm just as interested in why only two out of the five animals could do it (Prior, Schwarz, & Güntürkün, 2008). By conceding that individual differences do, in fact, exist and exert constraints on cognition (either by facilitating or limiting it), a dialogue can be opened up about why a behavior like self-recognition might be susceptible to the influence of individual differences, especially since it's fundamental to what humans would call a self-concept.

As the diversity of species that are studied increases, experts in the field have noted a renewed interest in thinking about individual differences and their impact on cognition (Shaw & Schmelz, 2017). This makes sense since the two are realistically inseparable.

PERSONALITY

In the human psychology world, personality is considered someone's enduring or stable, predictable ways of thinking, feeling, or behaving over time and across context. A personality is what makes your outlook on the world uniquely yours and different from mine. Given the historical significance, personality in animals is often referred to as "individual differences" because there is high variability between individuals but low variability within an individual. In other circles, rather than saying *personality*, scientists prefer "temperament" to reflect a more hardwired, genetic perspective. Biologists who study animal behavior also avoid saying *personality*, electing to call stability in responding an animal's "behavioral syndrome" instead. Regardless of the term used, everyone would agree that Pancho consistently and predictably behaved in ways that made us think he was grumpy and bitter at the world and that his disposition was different from other iguanas' for a number of reasons related to his unique biological makeup and life experiences.

Personality emerges out of a dynamic interaction between nature and nurture. Individuals can be genetically predisposed to certain features of their personality either by inheriting traits or by random mutation. Case studies of identical twins who were separated at birth illustrate just how strong of an influence genes have on our personality. Even when two individuals are raised in drastically different life circumstances by completely different families, their identical genetic makeup means at least some features of their personality will be shared. But genes are not everything. The environment *does* matter. Past history of trauma, repeated failure, or other salient life experience interact within the framework of biology to shape how individuals' personalities emerge. There's no reason to believe that these processes would not happen similarly in animals, albeit in different degrees based on species' complexity. Both dolphins and bees can learn—that is, their behavior can be affected by previous experience—however, dolphins can store memories for longer, show greater depth in their social interactions, and are more likely to have self-awareness. Thus, while we might predict a few basic personality traits in bees, dolphins might have more complex personalities. Since stable personality traits have been observed in species as surprising as spiders (Holbrook, Wright, & Pruitt, 2014), lizards (Carter, Goldizen, & Tromp, 2010), fish (Colleter & Brown, 2011; Stein, Trapp, & Bell, 2016), firebug larvae (*Pyrrhocoris apterus*, Gyuris, Feró, & Barta, 2012), and wild guinea pigs (*Cavia aperea*, Guenther, Finkemeier, & Trillmich, 2014), it would be unwise to write off species at face value.

While many of us intuitively know that animals have personalities, we must rely on science to tell us what it might look like and how to measure it. In humans and nonhumans, personality research follows a similar trajectory. First, researchers devise methods to collect data pertaining to the individuals' typical behavior. In animals, this is done using a number of clever tasks, which we'll get to shortly. Those behaviors, some of which are actually emotional states (e.g., anxiety), are then clustered into specific traits like boldness, aggressiveness, or openness to new experience. The resulting core traits are built up into models that can then be tested on other animals of the same species to increase the sample size and see how valid and reliable the model is across populations. Animal personality researchers can find themselves working at any stage of the sequence, and some run the full gamut by developing, testing, and replicating their own models. In this section, we'll highlight some of the findings and methodological considerations that must be made at each of these stages.

Animal cognition researchers who work at the earlier stages of the personality model sequence investigate a range of different kinds of species-wide and species-specific behaviors and methodologies to assess them. To say that an animal typically displays the trait of grumpiness as part of its personality suggests there's either a definition with specific indicator behaviors we recognize or a threshold that must be reached to qualify. The same goes for the constellation of other enduring ways of responding that comprise an animal's personality. As we'll explain, animal personality is typically measured using descriptive, qualitative methods. Some kind of stimulus or situation is introduced, and the animal's behavior in response to it is recorded and compared to what is known about an average member of that species. For example, one could observe the specific physical reactions an animal has in a given environmental context and categorize those physical responses as particular emotional states like fear, anxiety, optimism, or apathy. Interpreting emotional states could also include reliance on behaviors such as facial expressions like smiling or teeth baring or postures like leaning forward, recoiling, or even aggressing. When animals use facial expressions and posturing similar to how humans use them (as with the case of chimpanzees; Parr & Waller, 2006), inferring emotional states can be less daunting, but we're rarely that lucky. For example, a smile from a rhesus macaque is an aggressive gesture, not a happy or friendly one (de Waal & Luttrell, 1985). Further, what about when a single behavior is used in multiple contexts by a species? Brandishing an open mouth could convey aggression, play, or sexual excitement, which makes categorizing the animal's underlying emotional state a bit of a guessing game.

To demystify behavior, some personality researchers have begun correlating emotional states—anxiety, for example—with more physiological markers for them. Every major class of animals we've covered in this book secretes some kind of stress hormone. In insects and spiders, it's called octopamine; in reptiles, birds, amphibians, and rodents, it's corticosterone; and in most other mammals and fish, it's known by the familiar name cortisol. By exposing animals to different kinds of stimuli and situations and then comparing their stress hormone levels to the animal's own baseline levels or an established norm for the species, traits like anxiety, which can take the form of many different physical responses, can be more objectively studied.

To test the relationship between physiology and behavior in African elephants (*Loxodonta africana*), one group of researchers rated the animals on 23 different personality adjectives and then measured their cortisol levels. They found that individual differences at the behavioral level very much correlated with individual difference in cortisol levels (Grand, Kuhar, Leighty, Bettinger, & Laudenslager, 2012). Specifically, those whose personalities—based on outward behavior—had been determined to be more fearful had generally higher than average cortisol, while those with lower resting levels of cortisol tended to be rated as less aggressive and, surprisingly, less social. While this might seem counterintuitive, elephants are social animals. Highly anxious animals (i.e., those with higher resting cortisol levels) may be perceived as more social because they're more likely to seek out support and reassurance from others.

In addition to inferences made via physical and physiological differences, a third area that is gaining popularity in the study of animal personality is animal vocalizations. Species with large and diverse repertoires, like dolphins, often have complex vocalizations that can be used to infer emotional states and inform personality profiles. So far, vocalizations seem promising; however, cataloging vocalizations and correlating them to any one emotional state is difficult and highly time-consuming. As it is hopefully clear, neither physical, physiological, or vocal measurements are optimal, and in the case of collecting physiological or vocal markers, the logistics could render them unfeasible for some research teams. But taken together and with consideration of their potential flaws, they are certainly a place to start, especially when multiple methods can be combined to offer corroborating evidence (Kuczaj, Highfill, Makecha, & Eskelinen, 2012).

Comparatively speaking, linking behaviors to personality traits in humans and animals is equally difficult, but for very different reasons. For humans, identifying those individual differences comprising personality is typically done using models like the ones we highlight in the "Human Application" section. Because we humans tend to self-reflect, sometimes too much, on inventories like the Big Five (Costa & McCrae, 1992; McCrae & Costa, 1997), we might get caught up on a question like, "I insult people"—which actually appears on the inventory—and think to ourselves, "Well, do they mean in a malicious way? Accidentally? A playful way? I definitely do it in a playful way, but I'm not trying to hurt anyone." Soon, we find ourselves down the rabbit hole dissecting the questions to answer in a way that we feel best represents our personality, but we may, in fact, be completely off base. There's also the very real possibility of social desirability bias, where someone might respond no to the insult question, for example, because they're worried how the test results might be used or that they might somehow expose them to be a terrible person. Thankfully, because animals cannot self-report and do not seem to care what we think about them, these two issues related to translating behavior into traits are absent when we work with them. On the flip side, for as flawed as self-report can be, it also makes trying to study individual differences logistically easier. Without self-report, we have no choice but to develop nonverbal methods, which come with their own challenges.

Very likely, you could give examples of a few behaviors that would correspond to extraversion in humans. What about in a dog? If you spend enough time with dogs, maybe you could come up with something. But what about a koala, or even a hermit crab? What about measuring how long it takes a hermit crab (*Pagurus bernhardus*) to come out of its shell after being startled (e.g., Briffa, Rundle, & Fryer, 2008)? Is that extraversion or something else? What does extraversion look like for those species? The lesson here is that personality traits in humans do not always translate well to animals, and the animals we do think they translate nicely for are potentially fraught with anthropomorphism and human bias rather than objectivity. Additionally, you might be able to imagine rating a dog on Big Five personality traits like extraversion or openness to new experience, but what about on conscientiousness? As a result, while many tasks exist for inducing the behaviors we'd like to measure, only some of them seem to "work," so to speak. These kinds of issues can make those who study animal personality sometimes feel like they're fumbling in the dark. Every species is different, and traits like extraversion and conscientiousness are human labels created by humans to describe human behavior. In this way, researchers are often forced to rely on tasks and models that have only minimally undergone the necessary standardization and confirmation that is observed in human personality inventories like the five-factor model (Kaufman & Rosenthal, 2009). However, with enough repeated trials across different situations (personality should be enduring, after all), consistent patterns do emerge.

One very popular test to measure traits associated with what humans call openness to experience is the open-field test. Behaviors such as the amount of time it takes the animal to enter an open space, to interact with any novel objects in the field, the amount of time spent along the walls of the arena (i.e., thigmotaxis), how many times the animal rears up on its hind legs, and so forth can all be measured at once. While it might seem simplistic, the ecological relevance of the open field is relatively high. Walking out into an open space, being attracted to new things (i.e., neophilia), and standing up tall could all get an animal killed in the wild. Thus, for those who are willing to venture out, despite the potential for dangers, we might refer to them as scoring high for openness to experience.

While thigmotaxis and neophilia may not be typical personality characteristics in humans, these behaviors in animals represent a tendency (or lack thereof) to explore, which includes tolerating or being open to new experiences (Walsh & Cummins, 1976). However, not everyone agrees that the open-field test and its associated behavioral measures actually indicate exploration. In a recent cautionary tale with myna birds (*Acridotheres tristis*), researchers collected traditional data on their birds in the open-field test—latency to enter the space, duration of time spent on the ground, the speed with which exploration occurred—but when they analyzed their data, they found that behavior in an open-field may not be a clear-cut indicator of exploration, but rather shyness (Perals, Griffin, Bartomeus, & Sol, 2017). While it might seem like an unnecessary semantics complaint, realistically, exploration and shyness are not opposites, so the possibility that this common task is not measuring what we've thought it had been all these years, is somewhat overwhelming.

Sometimes issues have more to do with the behaviors used to infer traits than the task itself. In one study, when baboons were tested for the personality trait boldness by measuring both their alarm responses and the amount of time they spent exploring novel objects, a different problem emerged (Carter, Marshall, Heinsohn, & Cowlishaw, 2012). While both measurements were supposed to give an indicator of how bold each baboon was, the results conflicted with each other. That is, knowing how much time an individual spent exploring novel objects did not correlate with their alarm behavior as it should if both point to boldness. The researchers hypothesized that alarm behaviors were also measuring some degree of anxiety, which could explain the lack of consistency.

This is a good place to pause. Here's a question: How do you know you're measuring what you want to measure? Those who develop tests of cognitive abilities and individual differences must be sure they're measuring what they actually set out to measure. In this section so far, we've seen multiple examples of how carefully researchers must work to ensure that the labels (e.g., exploration) they assign to behaviors accurately reflect what they're seeing, that the tasks themselves correspond to those labels (e.g., open-field test measuring exploration), and that the labels say something about a particular core personality trait (e.g., exploration as an indicator of openness to experience). When a label or task does not measure the construct it's supposed to, we say the test has low construct validity. Issues related to construct validity are real concerns in research where constructs must be built from the ground up. This is why personality models and theories of intelligence (later in this chapter) can take years to confirm as reliable and credible. In fact, it is not just the study of individual differences where construct validity should be considered. Construct validity is hard to establish when you're testing animals on cognitive tasks as well (Völter, Tinklenberg, Call, & Seed, 2018).

CHAPTER 7 INDIVIDUAL DIFFERENCES BETWEEN ANIMALS

When it comes to individual differences, the best way to increase construct validity is to have several tasks that you predict measure the same construct. Rearing in and walking all the way to the center of open field might both indicate something about a rat's exploratory tendencies, for example. What would you conclude if the rat you're testing scored high for both? Most likely that both measurements are somehow related to one another and to the same overarching construct. That is, they appear to be valid measures of the same construct. It does not take much extrapolating to see how developing tasks that have high construct validity would take a long time, and once efforts to establish a measure as having high reliability (i.e., works consistently over time, across individuals, across populations) get added to the equation, it can take even longer. Nonetheless, to avoid confusion like what those studies described earlier experienced, attention to both construct validity and reliability must be given so that researchers give themselves the best possible chance to uncover predictable and consistent ways to study animal behavior.

Researchers developing the tasks and behaviors of interest that are used to correlate with specific personality traits face an uphill climb related to validity, semantics, and communication barrier. Like psychologists who study human personality, it can take years to refine the methods in order to arrive at personality traits that are both valid and consistent over time and across populations of the same species. From there, the core traits that seem to define a species' personality can be tested repeatedly, at which point it might be published as a model (e.g., five-factor model). There are a handful of standardized, species-specific trait models of personality out there for animals (e.g., Gosling & John, 1999; Grand et al., 2012). As you might expect, researchers are most confident about—from a scientific stance—models that have been created for primates and dogs, species with whom we share many of the same behaviors or spend a great deal of time, respectively. One of the earliest validated scales for assessing personality in rhesus macaque monkeys came from a collaboration between two primatologists: one of whom exclusively applied his findings to human child development and one human personality researcher (Chamove, Eysenck, & Harlow, 1972). With a deep knowledge of macaque behavior, a lab full of more than 150 monkeys, and a strong idea of how to measure personality, the team set up a series of social and solitary scenarios and then measured each monkey's response. Amount of social play, the extent to which they avoided an aggressive conspecific, how often they bit or grabbed conspecifics in an aggressive way, and how often they physically clung to conspecifics were just some of the target behaviors they measured. Those behaviors were then lumped into three basic factors, or traits—affiliative, hostile, and fearful—which they believed mapped onto popular human factors at the time of extraversion, psychoticism, and emotionality.

Turning our attention to our closest relatives, chimpanzees' personalities have been distilled down to five core traits: reactivity (measured by behaviors like aggressive displaying), openness (playfulness), agreeableness (considerate behaviors), dominance (sexual behavior and aggression), and extraversion (social grooming). Six similar core personality traits among bonobos—which are by nature less aggressive and more prosocial than chimpanzees—have also been found: assertiveness, conscientiousness, agreeableness, openness, attentiveness, and extraversion (Weiss, Inoue-Murayama, King, Adams, & Matsuzawa, 2012). You'll notice for both species of ape, these core traits show a striking resemblance to the Big Five human personality traits of openness, agreeableness, conscientiousness, extraversion, and emotional stability. Interestingly, we must look to a non-ape primate species for evidence of complete mapping onto all five of the Big Five traits. Iwanicki and Lehmann's (2015) assessment of personality identified all five in the common marmoset (*Callithrix jacchus*), a small South American monkey that is evolutionarily rather distant from humans.

Those who go the route of building up a personality model from scratch, moving from developing behaviors and tasks to organizing behaviors into core trait models are said to be working bottom-up. Some, instead, elect to work top-down and take a model that already exists and adapt its content for another species (Freeman et al., 2013). For example, Grand et al. (2012) adapted a preexisting personality model for chimpanzees to use with their elephants. There are inherent benefits and drawbacks to both ways of doing it, which is why certain practices are recommended. First, when possible, multiple individuals should be tested—as many as possible—and multiple individuals from different populations. This can help address any effects of the environment that are unique to one research location which might drive presumed relationships among personality traits. In addition to testing multiple animals, multiple raters who are well versed in the operational definitions for each behavior they're observing or trait they're evaluating should be involved. Multiple people rating behaviors or traits allows for interobserver reliability to be evaluated and diminishes experimenter error, thus increasing

CHAPTER 7 INDIVIDUAL DIFFERENCES BETWEEN ANIMALS

validity and reliability. Finally, if possible, animals should be tested multiple times across a variety of settings to really ensure that what is being measured as a personality trait is, in fact, stable and enduring.

While the majority of animal personality research is conducted by researchers in academia, some animal models of personality have gone mainstream. The Canine Behavioral Assessment and Research Questionnaire, or C-BARQ, was created and validated by a team of dog cognition experts. The assessment studies outward behaviors as a way to infer information about personality and another topic in this chapter, intelligence. Using a method similar to that which is used to develop human personality and intelligence tests, the C-BARQ uses a survey of nearly 70 questions about the behavior of pet dogs. Analysis of the C-BARQ results demonstrated there was a specific set of eight core factors that held consistent for any one particular dog but varied when multiple dogs were compared to each other—that is, personality-like traits. These factors were stranger fear, stranger aggression, owner aggression, general fear, fear/aggression specifically toward other dogs, trainability, attachment, and a tendency to engage in chasing behavior (Hsu & Serpell, 2003; Serpell & Hsu, 2005).

When the C-BARQ was ready, the research team decided to use a strategy human personality psychologists use and advertised widely for participants to further test their model. Citizen scientists (people who are not necessarily professionally trained in research but are eager to learn and help out) can test their dogs with the C-BARQ and contribute to a large and growing pool of data by accessing the website www.dognition.com. The scientists behind Dognition provide a list of tasks that dog owners can complete at home with their pets. These tasks include following where the owner has pointed, delaying retrieval of food when told "no," and locating a treat underneath one of two opaque cups (Stewart et al., 2015). When the scientists at Dognition analyzed the data from citizen scientists' dogs, the results were actually quite similar to results which had come from their laboratory, and that's good news! If dog owners are able to obtain reliable results at home by testing their own animals, we can test far more animals than we can if we have to bring them to a lab for testing, plus this hits on the practices of testing multiple animals across multiple settings, which are important for any good personality assessment tool.

The tests used to assess animal personality cannot tell the whole story. Gaps must be filled by considering the evolutionary value of stable, within-species differences. From a social stance, animals in groups take on different roles based on sex, dominance, and scarcity of resources, and sometimes those roles even change over time. Personality measurements have validated this. For example, in marmosets, older primates tend to be less bold, and the males specifically tend to be more agreeable, as do younger females (Koski et al., 2017). In a society with strict and often sex-based hierarchical societies, it makes sense that certain stable ways of behaving would be selected for in some, but not all, individuals in a group. There is evidence that there's a correlation between where an animal falls on the bold/shy continuum and its body shape and structure (Kern, Robinson, Gass, Godwin, & Langerhans, 2016). In the wild, a bolder meerkat might have both the body and the disposition to be a better sentry and guard for the group. Finally, in many ways, evolution might select for animals living in large societies to have a particular, unique way of being because it affords predictability in how different individuals will respond. A particular male dolphin could have an aggressive personality, but at least others know what they're getting from him. An individual whose behavior fluctuates too much is difficult to read and could jeopardize group safety and cohesion.

Along those lines, it's important to point out that while evolution might favor individual differences, there's a limit. An animal who is too shy or too bold, even if the trait is stable, could also jeopardize group safety or individual survival. For animals in human care, which is the vast majority of animals who participate in personality studies, individual differences in personality are free to range in greater extremes, since their care is monitored by humans. The question to consider as a critique of animal personality research is the extent to which what we see in human care mirrors their wild counterparts. If not, for a laboratory study to claim "monkeys have personality," and to conceptualize personality in the way that humans are free to vary, might be overly inflating what's actually going on for the species as a whole.

EMOTIONAL WELL-BEING

For animals in human care, which includes pets, research subjects, and animals at zoos and aquariums—all of which we have mentioned repeatedly throughout this book—another way that individual differences emerge is in variations in emotional well-being.

Researchers are relatively certain that at least some animals have subjective experiences, precursors to emotion, or full-blown emotions (e.g., Cabanac, 1999; Panksepp & Burgdorf, 2003). As such, in addition to personality's stable effect on behavior and cognition, stimuli in the environment can cause acute changes in an animal's emotional state, which in turn can impact how it thinks and behaves. While this may also be the case in the wild, most studies on emotional well-being in animals are focused on those in our care.

Determining whether the life experiences we create for animals promote emotional well-being is tricky, like most features of animal cognition. At least one widely circulated attempt to measure general well-being emphasized response to stress as an indicator of an animal's emotional well-being (Moberg, 1985). Technically, stress can be both positive and negative. Positive stress can be thought of as the general arousal experienced across a variety of different fun or exciting experiences. We all know negative stress—no explanation necessary. Like humans, animals also show individual variation in how they respond to stressors. Coping style and stress reactivity are two ways that researchers can measure that variation (Koolhaas, de Boer, Coppens, & Buwalda, 2010). Stable, within-species differences in how an individual responds to a situation (i.e., coping style, qualitative description) as well as the intensity of that response (i.e., stress reactivity, quantitative measure) should differ based on biological predispositions and previous life experience.

Animals already experiencing positive and negative emotional states are more likely to respond to new situations in ways that look like optimism and pessimism, respectively. We refer to this shift in subjective overall outlook cognitive bias, and we find this phenomenon in species from fruit flies to chimpanzees (Bateson & Nettle, 2015; Deakin, Mendl, Browne, Paul, & Hodge, 2018). In the task, a negative (or positive) stimulus would be repeatedly presented over and over in order to build up a pessimistic or optimistic outlook, and then a new, ambiguous stimulus is presented. A consistent positive or negative outlook either experimentally induced or due to less than ideal living circumstances correlates with changes in behavior such as greater fear or exploration of ambiguous stimuli, which could very well affect that animal's cognition (Clegg, Rödel, & Delfour, 2017; Mendl, Burman, Parker, & Paul, 2009).

In 2015 the C-Well, or Cetacean Welfare Assessment (Clegg, Borger-Turner, & Eskelinen, 2015), was developed based on measures in the Farm Animal Welfare Quality Assessment. The C-Well uses an assortment of criteria to assess animal welfare. The tool measures features of an animal's general health (body condition, hydration, weight), behavior (social behaviors, response to trainers, tactile interactions), and emotional state to provide an individual welfare score. The C-Well is just one of many examples of how animal caretakers have addressed the importance of being knowledgeable of their animals' emotional health.

Studying emotional well-being and other individual differences that exist is useful for both scientific and practical reasons. Scientifically speaking, referring to an animal as having emotional states which contribute to an overall state of being suggests animals experience the world subjectively. Furthermore, that subjectivity is important to consider when conducting animal research of all types. Taking into account emotional well-being and personality is also beneficial to those who care for animals when it comes to making choices that are in the animal's best interests (Grand et al., 2012). For example, if a rhinoceros is always laying down at peak visitor time at the zoo, collecting baseline cortisol data could be helpful to know if the animal is behaving that way because it's perpetually relaxed or perpetually stressed out. If it's the latter, this can inform everything from medical treatment to housing decisions to interaction policies in order to address the animals' emotional needs. Even the process of selecting a pet could be improved. The Humane Society of the United States has shown that creating individual profiles for shelter cats and matching them to adopters actually resulted in fewer returns (Weiss, 2007). Going to an appropriate home that was matched to the animal's unique qualities and getting to stay there permanently likely boosts emotional well-being in those animals as well, illustrating the valuable application of this work.

ANIMAL INTELLIGENCE

By strict standards, intelligence and cognition are very different, though they're often conflated. In some ways, this makes sense. An animal that can interact with its environment in complex and flexible ways like those we've covered so far in this book presumably needs "more smarts" in order to do it, and vice versa. Here's a helpful

CHAPTER 7 INDIVIDUAL DIFFERENCES BETWEEN ANIMALS

way to think about this. Set a timer for 1 minute, and write down all of the uses you can come up with for a spoon that *do not* involve eating. To get you started, a spoon could be used as a tiny shovel.

How did you do? Five? Ten? If someone told you they came up with 30 creative uses for a spoon, what might you think about them? Probably that they are really smart. This is why some want to lump cognition and intelligence together. Nevertheless, cognition and cognitive flexibility do not really tell us much about an animal's intelligence (e.g., Mikhalevich, Powell, & Logan, 2017). This could be due, in part, to how intelligence is being defined (i.e., someone's ability to identify 30 uses for a spoon will tell you nothing about their intelligence if intelligence is being defined as performance on a calculus test). It could also have to the fact that intelligence as a field of study in humans has a long and somewhat sordid history attached to it. There are few definitive answers about what intelligence actually is in humans, let alone in animals; however, there are two basic schools of thinking—that intelligence is either one factor or comprised of many factors.

The idea that intelligence is one thing was first advanced by Charles Spearman. Spearman called this thing g, for *general intelligence* (Spearman, 1904). The best way to think of g is a measure of overall "smartness"—g theory says if you're smart, you're smart (and if you're not, then you're not). Spearman thought that if you were good at one skill, task, or ability, you were likely good at many—for example, if you were good at math, you were likely good at science and literature, too. In the second half of the 20th century, what became known as Cattell–Horn–Carroll, or CHC, theory was developed (Carroll, 1993; Cattell, 1963; Horn, 1968). CHC theory broke down intelligence into two basic types: fluid intelligence, or problem-solving ability, and crystallized intelligence, or factual knowledge. Fluid intelligence has since been further divided to include constructs such as long-term memory, short-term memory, and visuiospatial ability. Lastly, in the 1980s, Robert Sternberg introduced Triarchic Theory of Intelligence (TTI), which divided intelligence into three areas—practical, analytical, and creative. These terms are referred to as "textbook learning," problem solving, and creative abilities, respectively (Sternberg, 1985). Sternberg's TTI was similar to CHC and other theories at the time in that it allowed for individual differences in abilities—something not well accounted for by g—but different in that it was much more practical. Among current human intelligence researchers, CHC and other theories that conceptualize intelligence as being more than a single thing are generally favored over *g* theory (Kaufman, 2009).

Why does this history lesson matter? Because if the jury is still out on how *human* intelligence works, it is *way out* on how animal intelligence works. Traditionally, intelligence in animals has been looked at from a very practical perspective—much more similar to TTI, although there has not been a direct comparison of the two.[2] Is the animal able to survive? Does it find itself an ecologically sound niche to thrive in? Does it reproduce? (Reader, Hager, & Laland, 2011; Reader & Laland, 2003). Charles Darwin (1872) himself even said, "Intelligence is based on how efficient a species became at doing the things they need to survive."

The actual empirical measurement of intelligence in animals is a much newer endeavor. It is generally being undertaken by scientists who study animal cognition—not human intelligence. Most of these researchers have based their work around *g* and the premise that it exists (Burkart, Schubiger, & Van Schaik, 2016; Deaner, Nunn, & van Schaik, 2000; Lee, 2007). Thus, this work attempts to find, not confirm, *g*'s existence relative to other potential theories of intelligence like CHC (Kaufman, Reynolds, & Kaufman, 2019; Shaw & Schmelz, 2017). These studies often test rats and mice on scaled-down tasks presumed to be associated with intelligence—for example, mazes, short-term memory, and minor tool manipulation.

While much of the work in animal intelligence has focused on *g*, there are several tests- mostly with primates—that do show evidence for multiple factors of intelligence, rather than *g* (Herrmann & Call, 2012; Reader et al., 2011). At least one research team has specifically identified CHC's factors of fluid intelligence, crystallized intelligence, and visual processing in primates (Kaufman et al., 2019), while another study showed three different intelligence domains—physical, spatial, and social—in chimpanzees (Herrmann, Hernández-Lloreda, Call, Hare, & Tomasello, 2010). Based on the tasks these two studies used, both determined there was no global "if you're smart, you're smart" where each subject performed roughly equivalently on all of the tasks. Instead, one subject might excel at tasks that require a good spatial sense—like remembering where things were hidden or following hidden items when their containers are rotated—but underperform on tasks that require good social

---

2. To our knowledge.

abilities, such as following another's gaze or using a pointing gesture to show intentions.

One primate intelligence test deserves specific mentioning because it appears that it is being used by multiple research groups, thus standardizing data better. In 2007, Herrmann and colleagues introduced the Primate Cognition Test Battery (PCTB; Herrmann, Call, Hernàndez-Lloreda, Hare, & Tomasello, 2007). The tasks in the PCTB are divided into physical and social sections, called domains, and placed into groups (called scales) by the abilities they measure. For example, within the physical domain, there are "space" and "causality" scales. The "space" scale involves tasks measuring object permanence or spatial memory, while the "causality" scale might require an animal to understand that shaking a container with an object in it makes noise. The PCTB is unique in that both its format—domains and scales—and the types of tasks it uses (object movement, numerical comparisons, etc.) are similar to the type of tasks that are used in human intelligence tests. As a result, the same statistical methods used to analyze human intelligence tests can also be used on PCTB scores (Kaufman et al., 2019). Interestingly, the most recent studies with the PCTB seem to produce results that fit better into a model with multiple types of intelligence (such as CHC), than a single g-like structure. (Herrmann et al., 2007, 2010; Hopkins, Russell, & Schaeffer, 2014).

When the tasks on the PCTB were given to children, Herrmann et al. (2007) obtained results that led to what they named the cultural intelligence hypothesis. Both humans and chimpanzees scored similarly on PCTB tasks in both physical and social domains until the children reached approximately 2 to 2.5 years of age. At this point, the children's scores on the social tasks increased dramatically. Herrmann et al. hypothesized that this was due to the appearance of what they termed "cultural intelligence" in the children—but not the chimpanzees. They further hypothesized that what they were seeing was a reflection of the impact of complex human social structure and cultural traditions on intelligence (Herrmann et al., 2007).

While the study of human intelligence dates back almost to psychology's inception, the measurement of intelligence in animals is still very much in its nascence. Given this, it will take time to build a solid repertoire of assessments and methods. If what we have learned so far is any indication, there is much to be discovered.

ANIMAL SPOTLIGHT: ALEX

Alex (Avian Learning Experiment), the African grey parrot who changed what we know about avian intelligence, became a research subject completely by chance. This was actually the intention Irene Pepperberg had when she walked into a Chicago pet store in 1977. The store had a group of year-old African grey parrots, and Pepperberg asked the manager to pick one at random for her (Pepperberg, 1999).

Pepperberg had a new idea about how to work with Alex—she was going to use the model-rival method to teach him (Todt, 1975). Todt developed the model-rival technique in response to the prevailing animal training method of using food as a reward. Instead, Todt tapped into parrots' highly social nature (Pepperberg, 1988, 1993; Trestman, 2015) and let a human assistant model correct responses. Over time, the animal learned what the correct responses were by observing the model, and voluntarily participating in the training session. Pepperberg's procedure with Alex was simple—the trainer would ask the assistant to label an object in front of Alex, and when the assistant correctly labeled the object (e.g., "truck" for a toy truck), the trainer would praise the assistant and give the object to them to interact with. Eventually, curious Alex began vying for the trainer's attention as well as the opportunity to interact with new objects. Pepperberg attributed her success to social learning. She was familiar with psychologist Albert Bandura's work on social learning in humans and the tendency for social animals like humans to soak up information from social situations, so she figured the same theories might apply to Alex. Bandura's work showed the significance to the learner of observing others' behavior and its consequences, in addition to how that behavior fits into the general social context in which it occurred (Bandura, 1966). This makes sense—my (AK) son is much faster to learn that playing baseball is not allowed in the house when he sees his brother get in trouble for it as opposed to when I tell him so.[3]

Alex began his learning with basic labels for color, shape, size, and "matter" (material). This foundation was necessary before any further work could be done. After about two years of training, Alex

---

3. That's assuming my boys have indeed learned that, which is debatable.

had mastered the names of nine items, along with three colors and two materials (e.g., wood). He could combine these labels (e.g., "blue paper") and eventually reached a point at which he was correct approximately 80% of the time (Pepperberg, 1999). Color is a particularly difficult construct because it's abstract—that is, I cannot show you "green." I can show you a green paper and a green key and identify them as such. But a certain level of flexible cognition is required first to identify what is common between the two items and, second, that it is represented by the word *green* (Pepperberg, 1992, 2006). Yet another cognitive step is required to understand that "green" can be transferred to new objects that were not part of the original training process, like being able to label a ball as green, if training had only ever involved green paper and keys.

In addition to learning labels and abstract concept categories, Alex also advanced our understanding of how parrots perceive quantities. He was able to answer "how many" questions for up to eight objects, but, more interestingly, he was able to identify quantities of subsets of items. For example, on a tray with keys and toy trucks, Alex could identify how many keys there were, and he could add a secondary label and provide the number of green keys. This is a particularly complex task—although it may not seem so at first (Pepperberg, 1992). To answer it correctly, Alex could not just count the number of keys or the number of green objects. Instead, Alex needed to identify all of the objects on the tray that overlapped as both green and keys, count those objects, and then articulate his response aloud using a communication system that is not natural to African greys.

Once he learned basic counting, Alex became conversant in the use of "zero"—again, a concept that might seem simple at first (Pepperberg & Gordon, 2005). Take a moment, and think about how you might teach someone with Alex's vocabulary what zero means. Zero is the absence of something. You could place an object on a tray and remove it and say there are now zero of that object, but it's far more likely your student will be more focused on the action of taking the object away than the current lack of it. The same challenge could be said for the relational concepts of "bigger" and "smaller" than—which are also abstract ideas Alex eventually mastered (Pepperberg, 1999).

One of the hallmark features of language is generativity, the ability to recombine a limited number of words in order to communicate infinite information. Linguist Noam Chomsky's (1957/2009) famous example for generativity, "Colorless green ideas sleep furiously"

(p. 15), illustrates this perfectly. Alex, too, could engage in a form of generativity by creating new labels for objects. One widely shared instance of this occurred when Alex requested a "banerry." The object of his request was an apple, but Alex had trouble saying the "p" sound. Although nothing can be said for sure, those who worked with him were very confident that he was combining two carefully selected words he could say: banana (which has a similar taste to an apple) and cherry (which has a similar appearance to an apple; Pepperberg, Brese, & Harris, 1991).

Pepperberg has several other parrots in her lab. She trained them all using the model-rival method and has demonstrated, for example, that they can use logic and reasoning in an assortment of linguistic and mathematical tasks (Clements, Gray, Gross, & Pepperberg, 2018; Pepperberg, Gray, Mody, Cornero & Carey, 2018). Still, while these other parrots have been successful, none have reached Alex's level of achievement. Why? What made Alex different?

The anticlimatic answer is, of course, we do not know. But just as people differ in skills and abilities, animals do as well. It also depends on whom you compare Alex to and how their life histories differ. Pepperberg once challenged Alex and three of her other companions to the famous bird problem-solving task, the string pull test. Two birds, Arthur and Kyaaro, immediately got to work and pulled up the string in the traditional sequenced fashion to reach their prize. Griffin and star student Alex, on the other hand, failed miserably despite multiple tries and encouragement (e.g., "Go pick up nut"). This is where life history, individual differences, and cognition matter. Griffin and Alex were Pepperberg's more verbal birds. Rather than toiling with the task, these two birds simply looked at the experimenter and said "Want nut." Again and again, they requested whatever had been tied to the bottom of the string instead of working for it. It appeared that their mastery of specific language-like skills had changed their entire outlook on the situation (Pepperberg, 2007)!

Moving out of the laboratory and into the field, how does Alex's behavior compare to a wild African grey's? In both cases, the raw materials are the same. A parrot brain is a parrot brain. The difference between Alex and a wild grey is the effect of the environment (Kaufman & Kaufman, 2016; Lloyd, 2004; Trestman, 2015). Compared to Alex, wild greys have to devote time and energy to meet their own survival needs, they live in environments with limited amounts of resources and experiences, they're reinforced for different kinds of behaviors (e.g., a wild grey is rewarded with nuts for disassembling tough husks, while destroying equipment is highly frowned upon in the laboratory), and the amount and kinds of social contact are very different. It's easy to see how an encultured parrot in human care could develop a distinctly different disposition from wild parrots and master complex cognitive tasks. Even in the wild, difficult problems requiring novel solutions arise, and there are individual differences in how they are addressed. For example, less dominant individuals tend to be better innovators than more dominant ones. Why? A dominant animal has to spend time and energy maintaining his or her status, defending territory, and generally looking out for the group. A subordinate animal does not have these kinds of distractions, and so the additional time and energy can be directed elsewhere (Reader & Laland, 2001). Still, no matter how much innovating wild parrots do in the wild, tests like those Alex encountered occur few and far between in nature, if at all in some cases. This makes what Alex could do all the more remarkable.

HUMAN APPLICATION: PERSONALITY TESTS

Confess, you've done a personality test or two. Or seven. You're a Blue Peacock. Or maybe an Owl. How about a Social Realist? Oh, and your dominant trait is Kindness. Are you an ISFJ or an ENTP? How well do the personality tests we fill out online or in magazines stack up to those psychologists develop, and what is the science behind measuring personality in humans?

The truth is that just about every personality test you've taken is not particularly scientific. It turns out your personality cannot really be measured in 10 questions! There are specific tests that psychologists use to measure personality—but generally they're not the fun ones you find online. Developing them takes many years of writing, trying it out, revising, and so on in order to hone in on written questions that capture personality. To explore why it's so challenging, let's first consider the person-situation controversy. Personality has many variables or factors. For example, generally people are not just shy; they're more likely shy around large groups of new people but outgoing in smaller groups or if they know everyone they're with. Personality is supposed to be enduring and stable, but as this example and multiple psychologists have pointed out, people do not always act predictably (Mischel, 2009), making it pretty tough to accurately measure or explain.

Another reason why personality is so difficult to study is that there are many different theories for how personality develops in humans—ranging from the impact of childhood experiences to biochemical predispositions. The main challenge personality psychologists face is that psychology wants to determine underlying principles that can then be applied to explain and predict behavior and mental processes generally. This directly conflicts with the fact that personality is what makes us unique and that it seems to be influenced by social context. How does someone develop a generalizable principle for the way uniqueness emerges and can be measured? This question has left psychologists scratching their heads for more than a century.

The personality theory we'll be focusing on is exemplified by the opening of this section: trait theory. It's also the one favored by those who study individual differences in animals. Trait theorists look for underlying ingredients, or traits, that are found in everyone to some degree. By boiling down personality to a handful of core traits that vary in unique combinations, trait theorists have developed generalizable principles for personality. One common trait model is the Big Five or five-factor model. According to this model, every human personality is composed of five core traits: conscientiousness, agreeableness, emotional stability, openness to experience, and extraversion (Digman, 1990; McCrae & John, 1992). Each trait is on a spectrum, meaning someone could be low on one factor and high on another. Low and high correspond to more or less visibility in an individual's personality. People who take the Big Five personality inventory have to decide how much statements like "I have excellent ideas" and "I get irritable sometimes" reflect themselves. While they might seem haphazard, the questions on the Big Five inventory were carefully developed by the consensus of years of research and hundreds of studies. Over and over again, across generations, geography, and participant age, the Big Five seems to hold up as a reliable measurement tool.

One of the most comprehensive personality inventories is the Minnesota Multiphasic Personality Inventory, or the MMPI (Hathaway & McKinley, 1940). It consists of 10 "scales" that examine constructs like depression, social introversion, mania, hysteria, psychopathy, and traditional gender roles. In order to accurately capture so much about someone's personality, the original full MMPI is gargantuan—nearly 600 true/false questions, including "I think I would enjoy the work of a librarian" and "My father is/was a good man." Unlike the five-factor model, whose main goal is just to measure personality, scores for the MMPI have also been used for decades as a clinical diagnostic tool by trained mental health service providers (Ben-Porath & Butcher, 1989).

While we might think we are good judges of our own personality, having a personality inventory administered by a professional can be very informative and highly applicable to daily life. Personality inventories are used in clinical and counseling settings, schools, and businesses to predict everything from management styles to how two people might improve their communication skills (Furnham, 2003; Myers & Myers, 1980; Pittenger, 2005). Developmental psychologists and school counselors may find it useful to understand why children of the same age may show different social or emotional stages, and social psychologists can use the traits to understand how lasting friendships are formed (McCrae & John, 1992). The MMPI specifically has been used in cases involving personal injury lawsuits as well as instances in which employers need job applicants who can handle working under stressful conditions, such as hostage negotiators, airline pilots, or law enforcement (Ben-Porath & Butcher, 1989).

Personality inventories can also be found in more informal settings, like a high school guidance counselor's office, where mentors can work with students to help them make decisions about their future. For example, according to the Myers–Briggs Type Indicator (MBTI; Myers & McCaulley, 1985), those who perceive the world intuitively (i.e., think more abstractly) tend to seek out change and might therefore find happiness in occupations like teachers or other "people-oriented" helping occupations. Given its wide-reaching use, you can take the MBTI for free online, but keep in mind, its scientific validity is low, and so your results should be considered with a critical eye. The MBTI scores test takers on dichotomies (i.e., either Intuitive *or* Sensing) rather than a spectrum, which means someone who scored 51% Intuitive would be boxed in as Intuitive rather than Sensing, despite the scores being basically an even split (Pittenger, 2005). There's also the issue of the person-situation controversy mentioned earlier. After 6 weeks of working as a leader at a sleepaway camp with hundreds of screaming kids—when all you can think about is your nice, quiet apartment—how do you think you'll score? Probably as an Introvert rather than Extrovert.

Heavily scrutinized and validated professional personality inventories like the five-factor model or MMPI differ drastically from personality tests we might find on the Internet and in magazines. Part of the difference is in the careful wording of questions, rigorous testing, and number of participants involved to really hone in on specific questions that can capture an individual's unique personality. On the other hand, most of the personality tests we encounter online and in magazines are so generically written that you could find truth in any of the results they tell you. Interestingly, there's science to back this up. The Barnum Effect[4] (also known as the Forer Effect) says we love to believe that vague information applies specifically to us, when really it can apply to anyone. Forer (1949) first identified this phenomenon by having his participants fill out a personality inventory then returning the exact same personality results to everyone a week later. When they were asked to evaluate the test's ability to capture their unique personality, all of the participants raved about it. Sadly, this means if the four possible personality types on an Internet personality test are Wise Wildebeest, Kind Kangaroo, Irritable Ibis, and Empathetic Elephant, according to the Barnum Effect, you might score Wise Wildebeest, but you're really a Gullible Gnu.

YOUR TURN!

Measuring individual differences in animals is already tough, so you can imagine what a time we had trying to come up with a one-size-fits-all, species-wide DIY ministudy for our final research-at-home idea. Here, we'll look at two dimensions: bold versus shy and novelty seeking. We'd expect individual differences to correlate with how important it is for that species to play it safe or need to explore in order to survive. You might already have a hypothesis of your own based on what you know about your chosen species' natural history. If you'll be working with a pet and you're feeling ambitious, you could also test multiple animals from the sample species from multiple households. Research shows that features of humans' personalities correlate with their pets' personalities (Dodman, Brown, & Serpell, 2018), so all kinds of fascinating individual differences might emerge!

---

4. Named for showman P. T. Barnum, who was known for sideshow advertisements that were just vague enough to make you curious about what was inside—the clickbait headlines of the 1800s.

To conduct this ministudy, you'll need a novel object that your animal has never seen before, preferably something that moves or is otherwise attention grabbing. Taking inspiration from Merola, Lazzaroni, Marshall-Pescini, and Prato-Previde's (2015) object from their cat social cognition study, you might use a small oscillating fan with a few short streamers tied to it. Now *that's* something no one sees every day. Whatever you use, go for "odd." You'll also need a food-motivated animal that you can interact with directly, some of that animal's favorite treats, a video camera, a stopwatch, and two small, identical barriers. The barriers could be opaque plastic cups if your research subject is able to knock them over to get treats underneath; two boxes that can be set up with the treats hidden behind them, making it easy for smaller or less agile animals to walk around them to get to the treat; or any other barriers that work for the needs of your research subject.

For the bold-versus-shy test, set up your novel object on the opposite side of the room from your research subject, and then have a seat by your animal and start filming. Working from a video recording will be easier than live recording your subject's behavior because of the nuanced things you might miss while you are taking notes. Start the stopwatch, and record for 5 minutes. If your research subject tries to interact with you, avoid engaging them, unless they seem genuinely afraid or stressed out. If this is the case, you can stop now because you've learned enough about your animal's personality in those first few seconds than is worth putting them through further unnecessary stress.

After 5 minutes, remove the object, and review your video footage. Look for general behaviors such as whether they approach the object at all and, if so, how long it takes them; whether they interact with the object and how; and whether they approach it playfully, aggressively, fearfully, or curiously. You'll need to develop operational definitions for these last terms since every species is different.

To test how your animal rates for novelty seeking, start with one of the barriers and a handful of treats. Repeatedly place a treat under (or behind, whatever is best for your subject) the barrier until your subject will reliably retrieve it. Play this game over and over and over again, making sure the barrier stays in the same location. After 15 successful retrievals in a row, prevent your subject from watching, and put up the second barrier next to the first. Continue to play the game of putting food under/behind the first barrier. Observe and make notes on your animal's behavior. Do they continue to only go for the original, reliable location for food and completely ignore the second? Do they test the second barrier to see what's under it? If so, after how long, and what is their behavior like? Does it seem hesitant, or does it seem like your subject has an expectation that there's something exciting to explore there, too?

All of the data you collect can give you an idea of your animal's personality, and you can try variations to see how your animal responds. Perhaps your subject was not willing to approach the novel object, but put a piece of food in front of it and they boldly approach to get it. This might indicate easily overcome cautiousness rather than shyness. The more you play around with this setup, the more you may be inspired to create another task or to test more of the same (or different) species, and the more you'll learn.

REFERENCES

Bandura, A. (1966). Observational learning as a function of symbolizationand incentive set. *Child Developement, 37*(3), 499–506. doi:10.2307/1126674

Bateson, M., & Nettle, D. (2015). Development of a cognitive bias methodology for measuring low mood in chimpanzees. *PeerJ, 3*, e998. doi:10.7717/peerj.998

Ben-Porath, Y. S., & Butcher, J. N. (1989). The comparability of MMPI and MMPI-2 scales and profiles. *Psychological Assessment: A Journal of Consulting and Clinical Psychology, 1*(4), 345–347. doi:10.1037/1040-3590.1.4.345

Briffa, M., Rundle, S. D., & Fryer, A. (2008). Comparing the strength of behavioural plasticity and consistency across situations: Animal personalities in the hermit crab Pagurus bernhardus. *Proceedings Biological Sciences, 275*(1640), 1305–1311. doi:10.1098/rspb.2008.0025

Burkart, J. M., Schubiger, M. N., & Van Schaik, C. P. (2016). The evolution of general intelligence. *Behavioral and Brain Sciences, 40*, e195. doi:10.1017/S0140525X16000959

Cabanac, M. (1999). Emotion and phylogeny. *Japanese Journal of Physiology, 49*, 1–10. doi:10.2170/jjphysiol.49.1

Carroll, J. B. (1993). *Human cognitive abilities: A survey of factor analytic studies.* New York, NY: Cambridge University Press.

Carter, A. J., Goldizen, A. W., & Tromp, S. A. (2010). Agamas exhibit behavioral syndromes: Bolder males bask and feed more but may suffer higher predation. *Behavioral Ecology, 21*(3), 655–661. doi:10.1093/beheco/arq036

Carter, A. J., Marshall, H. H., Heinsohn, R., & Cowlishaw, G. (2012). How not to measure boldness: Novel object and antipredator responses are not the same in wild baboons. *Animal Behaviour, 84*(3), 609–603. doi:10.1016/j.anbehav.2012.06.015

Cattell, R. B. (1963). Theory of fluid and crystallized intelligence: A critical experiment. *Journal of Educational Psychology, 54*, 1–22. doi:10.1037/h0046743

Chamove, A. S., Eysenck, H. J., & Harlow, H. F. (1972). Personality in monkeys: Factor analyses of Rhesus social behavior. *Quarterly Journal of Experimental Psychology, 24*, 496–504. doi:10.1080/14640747208400309

Clegg, I. L. K., Borger-Turner, J. L., & Eskelinen, H. C. (2015). C-Well: The development of a welfare assessment index for captive bottlenose dolphins (*Tursiops truncatus*). *Animal Welfare, 24*(3), 267–282. doi:10.7120/09627286.24.3.267

Clegg, I. L. K., Rödel, H. G., & Delfour, F. (2017). Bottlenose dolphins engaging in more social affiliative behaviour judge ambiguous cues more optimistically. *Behavioural Brain Research, 322*, 115–122. doi:10.1016/j.bbr.2017.01.026

Clements, K. A., Gray, S. L., Gross, B., & Pepperberg, I. M. (2018). Initial evidence for probabilistic reasoning in a grey parrot (*Psittacus erithacus*). *Journal of Comparative Psychology, 132*, 166–177. doi:10.1037/com0000106

Colleter, M., & Brown, C. R. (2011). Personality traits predict hierarchy rank in male rainbowfish social groups. *Animal Behaviour, 81*(6), 1231–1237. Retrieved from http://www.sciencedirect.com/science/article/pii/S0003347211001084

Costa, P. T., & McCrae, R. R. (1992). Normal personality assessment in clinical practice: The NEO personality inventory. *Psychological Assessment, 4*(1), 5–13. doi:10.1037/1040-3590.4.1.5

Darwin, C. (1872). *Expression of the emotions in man and animals*. London, UK: John Murray.

de Waal, F. B. M., & Luttrell, L. M. (1985). The formal hierarchy of rhesus macaques: An investigation of the bared-teeth display. *American Journal of Primatology, 9*(2), 73–85. doi:10.1002/ajp.1350090202

Deakin, A., Mendl, M., Browne, W. J., Paul, E. S., & Hodge, J. J. L. (2018). State-dependent judgement bias in *Drosophila*: Evidence for evolutionarily primitive affective processes. *Biology Letters, 14*, 20170779. doi:10.1098/rsbl.2017.0779

Deaner, R. O., Nunn, C. L., & van Schaik, C. P. (2000). Comparative testing of primate cognition: Different scaling methods produce different results. *Brain, Behavior, and Evolution, 55*, 44–52. doi:10.1159/000006641

Digman, J. M. (1990). Personality structure: Emergence of the Five-Factor Model. *Annual Review of Psychology, 41*(1), 417–440. doi:10.1146/annurev.ps.41.020190.002221

Forer, B. R. (1949). The fallacy of personal validation: A classroom demonstration of gullibility. *Journal of Abnormal and Social Psychology, 44* (1), 118–123. doi:10.1037/h0059240

Freeman, H. D., Brosnan, S. F., Hopper, L. M., Lambeth, S. P., Schapiro, S. J., & Gosling, S. D. (2013). Developing a comprehensive and comparative questionnaire for measuring personality in chimpanzees using a simultaneous top-down/bottom-up design. *American Journal of Primatology, 75*(10), 1042–1053. doi:10.1002/ajp.22168

Furnham, A., Moutafi, J., & Crump, J. (2003). The relationship between the revised NEO-personality inventory and the Myers-Briggs type indicator. *Social Behavior and Personality, 31*, 577–584. doi:10.2224/sbp.2003.31.6.577

Gosling, S. D., & John, O. P. (1999). Personality dimensions in nonhuman animals. *Current Directions in Psychological Science, 8*(3), 69–75. doi:10.1111/1467-8721.00017

Grand, A. P., Kuhar, C. W., Leighty, K. A., Bettinger, T. L., & Laudenslager, M. L. (2012). Using personality ratings and cortisol to characterize individual differences in African elephants (*Loxodonta africana*). *Applied Animal Behaviour Science, 142*, 69–75. doi:10.1016/j.applanim.2012.09.002

Griffin, A. S., Tebbich, S., & Bugnyar, T. (2017). Animal cognition in a human-dominated world. *Animal Cognition, 20*, 1–6. doi:10.1007/s10071-016-1051-9

Guenther, A., Finkemeier, M.-A., & Trillmich, F. (2014). The ontogeny of personality in the wild guinea pig. *Animal Behaviour, 90*, 131–139. doi:10.1016/j.anbehav.2014.01.032

Gyuris, E., Feró, O., & Barta, Z. (2012). Personality traits across ontogeny in firebugs, Pyrrhocoris apterus. *Animal Behaviour, 84*(1), 103–109. Retrieved from http://www.sciencedirect.com/science/article/pii/S0003347212001728

Hathaway, S. R., & McKinley, J. C. (1940). A multiphasic personality schedule (Minnesota) : I. Construction of the schedule. *The Journal of Psychology, 10*(2), 249–254. doi:10.1080/00223980.1940.9917000

Herrmann, E., & Call, J. (2012). Are there geniuses among the apes? *Philosophical Transactions of the Royal Society of London. Series B, Biological Sciences, 367*(1603), 2753–2761. doi:10.1098/rstb.2012.0191

Herrmann, E., Call, J., Hernàndez-Lloreda, M. V., Hare, B. A., & Tomasello, M. (2007). Humans have evolved specialized skills of social cognition: The cultural intelligence hypothesis. *Science, 317*(5843), 1360–1366. doi:10.1126/science.1146282

Herrmann, E., Hernández-Lloreda, M. V., Call, J., Hare, B. A., & Tomasello, M. (2010). The structure of individual differences in the cognitive abilities of children and chimpanzees. *Psychological Science, 21*(1), 102–110. doi:10.1177/0956797609356511

Holbrook, C. T., Wright, C. M., & Pruitt, J. N. (2014). Individual differences in personality and behavioural plasticity facilitate division of labour in social spider colonies. *Animal Behaviour, 97*, 177–183. doi:10.1016/j.anbehav.2014.09.015

Hopkins, W. D., Russell, J. L., & Schaeffer, J. (2014). Chimpanzee intelligence is heritable. *Current Biology, 24*(14), 1649–1652. doi:10.1016/j.cub.2014.05.076

Horn, J. L. (1968). Organization of abilities and the development of intelligence. *Psychological Review, 75*, 242–259. doi:10.1037/h0025662

Hsu, Y., & Serpell, J. A. (2003). Development and validation of a questionnaire for measuring behavior and temperament traits in pet dogs. *Journal of the American Veterinary Medical Association, 223*(9), 1293–1300. doi:10.2460/javma.2003.223.1293

Iwanicki, S., & Lehmann, J. (2015). Behavioral and trait rating assessments of personality in common marmosets (*Callithrix jacchus*). *Journal of Comparative Psychology, 129*(3), 205–217. doi:10.1037/a0039318

Kaufman, A. B., Reynolds, M. R., & Kaufman, A. S. (2019). The structure of ape (*hominoidea*) intelligence. *Journal of Comparative Psychology, 33*(1), 92–105. doi:10.1037/com0000136

Kaufman, A. B., & Rosenthal, R. (2009). Can you believe my eyes? The importance of interobserver reliability statistics in observations of animal behaviour. *Animal Behaviour, 78*(6), 1487–1491. doi:10.1016/j.anbehav.2009.09.014

Kaufman, A. S. (2009). *IQ Testing 101*. New York, NY: Springer Publishing Company.

Kaufman, J. C., & Kaufman, A. B. (2016). Capacity, potential, and ability: Integrating different approaches to studying animal vs human creative processes. *RUDN Journal of Psychology and Pedagogics, 4*, 29–36. doi:10.22363/2313-1683-2016-4-29-36

Kern, E. M. A., Robinson, D., Gass, E., Godwin, J., & Langerhans, R. B. (2016). Correlated evolution of personality, morphology and performance. *Animal Behaviour, 117*, 79–86. doi:10.1016/j.anbehav.2016.04.007

Koolhaas, J. M., De Boer, S. F., Coppens, C. M., & Buwalda, B. (2010). Neuroendocrinology of coping styles: Towards understanding the biology of individual variation. *Frontiers in Neuroendocrinology, 31*, 307–321. doi:10.1016/j.yfrne.2010.04.001

Koski, S. E., Buchanan-Smith, H. M., Ash, H., Burkart, J. M., Bugnyar, T., & Weiss, A. (2017). Common marmoset (*Callithrix jacchus*) personality. *Journal of Comparative Psychology, 131*(4), 326–336. doi:10.1037/com0000089

Kuczaj, S., Highfill, L. E., Makecha, R. N., & Eskelinen, H. C. (2012). Why do dolphins smile? A comparative perspective on dolphin emotions and emotional expressions. In S. Wantanabe & S. A. Kuczaj (Eds.), *Emotions of Animals and Humans: Comparative Persepectives* (pp. 63–85). Tokyo: Springer Japan. doi:10.1007/978-4-431-54123-3

Lee, J. J. (2007). A g beyond Homo sapiens? Some hints and suggestions. *Intelligence, 35*(3), 253–265. Retrieved from http://www.sciencedirect.com/science/article/B6W4M-4KXVD0C-1/2/3e26616a9947020e0525f95951c5da32

Lloyd, E. (2004). Kanzi, evolution, and language. *Biology & Philosophy, 19*(4), 577–588. doi:10.1007/sBIPH-004-0525-3

McCrae, R. R., & Costa, P. T. (1997). Personality trait structure as a human universal. *American Psychologist, 52*, 509–516. doi:10.1037//0003-066X.52.5.509

McCrae, R. R., & John, O. P. (1992). An introduction to the five-factor model and its applications. *Journal of Personality, 60*(2), 175–215. Retrieved from http://proxy.library.adelaide.edu.au/login?url=http://search.ebscohost.com/login.aspx?direct=true&db=pbh&AN=9208170743&site=ehost-live&scope=site

Mendl, M., Burman, O. H. P., Parker, R. M. A., & Paul, E. S. (2009). Cognitive bias as an indicator of animal emotion and welfare: Emerging evidence and underlying mechanisms. *Applied Animal Behaviour Science, 118*(3–4), 161–181. doi:10.1016/j.applanim.2009.02.023

Mischel, W. (2009). From personality and assessment (1968) to personality science, 2009. *Journal of Research in Personality, 43*, 282–290. doi:10.1016/j.jrp.2008.12.037

Merola, I., Lazzaroni, M., Marshall-Pescini, S., & Prato-Previde, E. (2015). Social referencing and cat–human communication. *Animal Cognition, 18*, 639–648. doi:10.1007/s10071-014-0832-2

Mikhalevich, I., Powell, R., & Logan, C. (2017). Is behavioural flexibility evidence of cognitive complexity? How evolution can inform comparative cognition. *Interface Focus, 7*(3), 20160121. doi:10.1098/rsfs.2016.0121

Moberg, G. P. (1985). Biological response to stress: Key to assessment of animal well-being? In *Animal stress* (pp. 27–49). New York, NY: Springer.

Myers, I. B., & Myers, P. B. (1980). *Gifts differing: Understanding personality type*. Mountain View, CA: Davies-Black Publishing.

Myers, I. B., & McCaulley, M. H. (1985). *Manual: A guide to the development and use of the Myers-Bnggs Type Indicator*. Palo Alto, CA: Palo Alto Consulting Psychologists Press.

Panksepp, J., & Burgdorf, J. (2003). "Laughing" rats and the evolutionary antecedents of human joy? *Physiology and Behavior, 79*, 533–547. doi:10.1016/S0031-9384(03)00159-8

Parr, L. A., & Waller, B. M. (2006). Understanding chimpanzee facial expression: Insights into the evolution of communication. *Social Cognitive and Affective Neuroscience*. Oxford, UK: Oxford University Press. doi:10.1093/scan/nsl031

Pepperberg, I. M. (1988). An interactive modeling technique for acquisition of communication skills separation of labeling and requesting in a psittacine subject. *Applied Psycholinguistics, 9*(1), 59–76. Retrieved from http://libproxy.lib.csusb.edu/login?url=http://search.ebscohost.com/login.aspx?direct=true&db=boh&AN=BACD198886020385&site=ehost-live

Pepperberg, I. M. (1992). Proficient performance of a conjunctive, recursive task by an African Grey parrot (*Psittacus erithacus*). *Journal of Comparative Psychology, 106*(3), 295–305. doi:10.1037/0735-7036.106.3.295

Pepperberg, I. M. (1993). A review of the effects of social interaction on vocal learning in african grey parrots (*Psittacus erithacus*). *Netherlands Journal of Zoology, 43*(1–2), 104–124. doi:10.1163/156854293X00241

Pepperberg, I. M. (1999). *The Alex studies: Cognitive and communicative abilities of grey parrots*. Cambridge, MA: Harvard University Press.

Pepperberg, I. M. (2006). Grey parrot numerical competence: A review. *Animal Cognition, V9*(4), 377–391. doi:10.1007/s10071-006-0034-7

Pepperberg, I. M. (2007). Individual differences in grey parrots (*Psittacus erithacus*): Effects of training. *Journal of Ornithology, 148*(Supplement 2), 161–168. doi:10.1007/s10336-007-0162-0

Pepperberg, I. M., Brese, K. J., & Harris, B. J. (1991). Solitary sound play during acquisition of English vocalizations by an African Grey parrot (*Psittacus erithacus*): Possible parallels with children's monologue speech. *Applied Psycholinguistics, 12*(02), 151–178. Retrieved from http://journals.cambridge.org/abstract_S0142716400009127

Pepperberg, I. M., & Gordon, J. D. (2005). Number comprehension by a grey parrot (*Psittacus erithacus*), including a zero-like concept. *Journal of Comparative Psychology, 119*(2), 197–209. doi:10.1037/0735-7036.119.2.197

Pepperberg, I. M., Gray, S. L., Mody, S., Cornero, F. M., & Carey, S. (2018). Logical reasoning by a Grey parrot? A case study of the disjunctive syllogism. *Behaviour, 1*, 1–37. doi:10.1163/1568539X-00003528

Perals, D., Griffin, A. S., Bartomeus, I., & Sol, D. (2017). Revisiting the open-field test: What does it really tell us about animal personality? *Animal Behaviour, 123*, 69–79. doi:10.1016/j.anbehav.2016.10.006

Pittenger, D. J. (2005). Cautionary comments regarding the Myers-Briggs Type Indicator. *Consulting Psychology Journal: Practice and Research, 57*(3), 210–221. doi:10.1037/1065-9293.57.3.210

Prior, H., Schwarz, A., & Güntürkün, O. (2008). Mirror-induced behavior in the magpie (*Pica pica*): Evidence for self-recognition. *PLoS Biology, 6*, 1642–1650. doi:10.1371/journal.pbio.0060202

Reader, S. M., Hager, Y., & Laland, K. N. (2011). The evolution of primate general and cultural intelligence. *Philosophical Transactions of the Royal Society of London. Series B, Biological Sciences, 366*(1567), 1017–1027. doi:10.1098/rstb.2010.0342

Reader, S. M., & Laland, K. N. (2003). Animal innovation: An introduction. In S. M. Reader & K. N. Laland (Eds.), *Animal innovation* (pp. 4–35). Oxford, UK: Oxford University Press.

Reader, S. M. M., & Laland, K. N. N. (2001). Primate innovation: Sex, age, and social rank differences. *International Journal of Primatology, 22*(5), 787–805. Retrieved from http://www.springerlink.com/index/ln47117526184421.pdf

Serpell, J. A., & Hsu, Y. (2005). Effects of breed, sex, and neuter status on trainability in dogs. *Anthrozoos, 18*(3), 196–207. doi:10.2752/089279305785594135

Shaw, R. C., & Schmelz, M. (2017). Cognitive test batteries in animal cognition research: Evaluating the past, present and future of comparative psychometrics. *Animal Cognition, 20*(6), 1003–1018. doi:10.1007/s10071-017-1135-1

Spearman, C. (1904). "General Intelligence," objectively determined and measured. *The American Journal of Psychology, 15*(2), 201. doi:10.2307/1412107

Stein, L. R., Trapp, R. M., & Bell, A. M. (2016). Do reproduction and parenting influence personality traits? Insights from threespine stickleback. *Animal Behaviour, 112*, 247–254. doi:10.1016/j.anbehav.2015.12.002

Sternberg, R. J. (1985). Implicit theories of intelligence, creativity, and wisdom. *Journal of Personality and Social Psychology, 49*, 607–627. doi:10.1037/0022-3514.49.3.607

Stewart, L., MacLean, E. L., Ivy, D., Woods, V., Cohen, E., Rodriguez, K., ... Hare, B. (2015). Citizen science as a new tool in dog cognition research. *PLoS One, 10*(9), 1–16. doi:10.1371/journal.pone.0135176

Todt, D. (1975). Social learning of vocal patterns and modes of their application in grey parrots (*Psittacus erithacus*). *Zeitschrift Fur Tierpsychologie, 39*, 178–188. doi:10.1111/j.1439-0310.1975.tb00907.x

Trestman, M. (2015). Clever Hans, Alex the Parrot, and Kanzi: What can exceptional animal learning teach us about human cognitive evolution? *Biological Theory, 10*(1), 86–99. doi:10.1007/s13752-014-0199-2

Völter, C. J., Tinklenberg, B., Call, J., & Seed, A. M. (2018). Comparative psychometrics: Establishing what differs is central to understanding what evolves. *Philosophical Transactions of the Royal Society B: Biological Sciences, 373*(1756). doi:10.1098/rstb.2017.0283

Walsh, R. N., & Cummins, R. A. (1976). The open-field test: A critical review. *Psychological Bulletin, 83*(3), 482–504. doi:10.1037/0033-2909.83.3.482

Weiss, A., Inoue-Murayama, M., King, J. E., Adams, M. J., & Matsuzawa, T. (2012). All too human? Chimpanzee and orang-utan personalities are not anthropomorphic projections. *Animal Behaviour, 83*(6), 1365–1355. doi:10.1016/j.anbehav.2012.02.024

Weiss, E. (2007). New research helps adopters meet their feline soul mates. In *Animal sheltering* (pp. 61–63). Washington, DC: Humane Society of the United States.

Conclusion

The rise of social media and popular press articles on the Internet has done wonders to make the field of animal cognition more visible and accessible to those outside of academia. Video clips like Betty the crow using tools are picked up and shared widely across social networks. Other media outlets are jumping on the animal cognition bandwagon as well. Entire issues of magazines like *Time* and *National Geographic* have been recently devoted to sharing what we have learned about animals' cognitive abilities. Animal cognition documentaries, like NOVA's *How Smart Are Animals?*, are increasing in popularity. In 2014, 12.6% of the 650 colleges and universities listed in *Forbes Magazine* offered a course focusing on animals' cognitive abilities (Abramson, 2015), and that number has likely increased since then. Courses on animal law are also being taught with greater frequency, which has paved the way for unprecedented legal actions like coparenting plans for dogs stuck in the middle of divorces, as well as arguments for legal personhood for captive chimpanzees.

In addition to animal cognition findings being used to change how we legally treat and view animals, the field of animal cognition has ushered in an exciting shift in the value of the work. People really want to know more about what other animals can do, which was one of the reasons we chose to write this book—and likely one of the reasons you chose to read it! In many ways, society has returned to the time of Clever Hans, where curious crowds gathered and followed Wilhelm von Osten and Hans across the country just to watch Hans count and tell time. Except this time, we have rigorous testing and methods to support our conclusions about the animal mind.

The science behind the field is constantly being updated to reflect new ideas, technology, and foci. We began this book with a crash course in the history of the field, and the West African sankofa bird's wisdom to look back to the past to know where you're going. Reflecting upon the past and present, we offer here some of our and other experts' observations and speculations about the future of animal cognition. In some cases, the predictions represent linear progressions; in others, the future direction is actually a circling back to old ideas. Because, let's face it, is anything *really* new anymore?

TECHNOLOGICAL ADVANCEMENTS

One of the biggest challenges that the field faced at its inception over a century ago has not and will not be going away anytime soon as far as we're concerned. Today, it's pretty clear that at least some animals have minds—though, we hope your inner behaviorist cringed at the idea of such certainty about something we cannot see or ask about. Herein lies the persistent challenge for the future. Even with all of the evidence presented in this book and beyond, we are still mainly drawing inferences about internal processes using outward behavior as a marker. As we have seen, an outward behavior that may look particularly complex, such as a horse that can do math or tell time, runs the risk of being misinterpreted. This is especially true since it's impossible to ask the animal to self-report or introspect about their experience. Imagine asking one of the ants in Cammaerts and Cammaerts' (2015) mirror mark study to explain what it sees when it looks in the mirror at the blue dot on its head. We'd give anything to have that existential conversation! For now at least, we must be content to make careful observations and even more careful inferences about what we're seeing.

Currently, we do not have an ant-to-human translator, and the possibility of ever developing one is more like something out of a

CHAPTER 8 CONCLUSION

science fiction novel than reality. On the other hand, neuroscience technology and methods like transcranial magnetic stimulation, positron emission tomography, and diffusion tensor imaging are being refined every day. These advancements let us access the inner workings of the brains of different species in order to test our behavioral inferences. One booming area of neuroscience comes from the canine cognition literature, where multiple laboratories have begun using brain scanning technology to observe how dogs' brains respond to stimuli in real time. MRI involves using very strong magnets to create high-resolution images of parts of the body. MRI's fancy cousin, fMRI, adds function to the structural images. In this way, researchers can present someone with images, such as faces, for example, and observe in real time how active a particular area of the brain is.

As it turns out, with lots of coaxing, treats, and mutual patience, dogs can be trained to stay inside the very loud, confining MRI tube long enough to have their brains scanned during tasks. The results so far have been an amazing window into just how special the dog–human bond really is. For example, if you've ever given your pet an old sweater to comfort them, you may have been onto something. Berns, Brooks, and Spivak (2015) tested what happens when dogs in an MRI tube are presented with the scent of themselves, a familiar human, strange human, familiar dog, or strange dog. As predicted, all of the scents made the areas of the brain associated with smell become active. What was not predicted, however, was that the only scent to activate the caudate nucleus—an area of the brain associated with reward and pleasure—was the familiar human. The fact that members of its own species, familiar or not, did not activate their reward center, and neither did the smell of a strange human, suggests the undeniable strength of the relationship dogs have with their owners. Now, we cannot conclude your dog *loves* you the way you might love your dog, especially since the reward activation your dog experiences when he smells you could be due to the fact you have been associated with food, but Berns et al.'s fMRI findings do show us that a dog's excitement about its owner goes beyond simple arousal and approaches something more akin to happiness.

Other cutting-edge areas of neuroscience include our deeper understanding of the mirror neuron system and its possible connection to complex social behavior. Mirror neurons are activated both when we complete a motor action ourselves, like grabbing a cup, and also when we watch someone else do it. Some scientists believe that mirror neurons are at the root of empathy or theory of mind—that they provide us with a way to connect with someone else engaging in an activity. By extension, this would mean connecting with someone else's thinking and the idea that everyone has their own individual thoughts (Agnew, Bhakoo, & Puri, 2007). As of now, the presence of mirror neurons has only been verified in monkeys and songbirds, in part because the methods used to locate and study them in the brain are highly invasive. In the near future, however, this may change. If it does, the connections between the mirror neuron system and complex sociality could be explored more broadly and more easily. In the future, we researchers could measure mirror neuron density in the brains of different species with varying levels of sociality as a core measurement in the quest to understand the evolution of complex social behavior.

Neuroscience was in its infancy when Margaret Floy Washburn published *The Animal Mind* in 1908. In fact, the first method for measuring brain activity, the electroencephalograph, was not invented until more than a decade later, and the MRI would not come until the 1970s. As both invasive and noninvasive neuroscience techniques advanced during the late 20th century, animal cognition researchers realized they could collaborate with neuroscientists in order to cross-check their behavioral findings with evidence from the inner workings of their subjects' brains. As time goes on, we can only imagine how processes like self-awareness, metacognition, theory of mind, and insightful problem solving can be better understood through neuroscientific methods.

INTERDISCIPLINARY COLLABORATIONS

Not everyone in animal cognition's history has subscribed to the practice of a diverse interdisciplinary perspective, and especially not psychologists who were trained and working during the behaviorist movement. Now, we train and collaborate in more interdisciplinary ways. Over the last few decades, the field has become more holistically informed and well rounded with respect to how studies are carried out and findings interpreted. Having a greater appreciation for the influence of an organism's natural history (i.e., biology, evolution, ecology) has helped psychologists, in particular, to better frame their findings in a more animal-centric, less sterile way. Gone are the days of Skinner's misinformed assumption that any animal could be trained to perform any behavior. Breland and Breland (1961), with their careful observations and thoughtful consideration of natural history, proved otherwise long ago. Similarly, biologists who spend time learning about the methodological blunders early psychologists made, such as the importance of experimental control of variables and developing appropriate tasks, can think more critically as they study animals' cognitive abilities in the field. With a better understanding of the species we work with, and the inherent methodological challenges, the quality, depth, and usefulness of our findings are enhanced.

Greater awareness of perspectives coming out of experimental psychology and biology critically offers a more accurate understanding of the organism. Interdisciplinary work asks not just whether a species has a particular ability but how it happens and why it might have evolved. It also challenges psychologists to step away from the traditional laboratory species like primates and rodents and to instead include a greater diversity of species, including the unexpected ones (Shaw & Schmelz, 2017). Fitting the pieces together can also help dispel myths about "intelligence" and hierarchies in cognition, as our evolutionary biologist friends teach us.

In our pursuit to develop a more interdisciplinary understanding of animal cognition, one outcome has been an increase in concern about the ethics of our interactions with animals (Kulick, 2017). For example, countries like Argentina, India, and Switzerland have ruled at least some animals are more than legal "things." In 2015, for example, an orangutan at the Buenos Aires Zoo named Sandra was legally determined to be a "nonhuman person." Citing sentience and intelligence in their argument, the Association for Assessment and Accreditation of Laboratory Animal Care (AAALAC) now strongly encourages all research with cephalopods to be passed through an ethics board first (AAALAC, n.d.). This is a particularly big deal since invertebrates have been historically exempt from research ethics board review. In some countries, our understanding of animals' complex social lives has even informed pet policies. For example, in Switzerland, it's illegal to own one guinea pig, and pet parrots are considered abused if they are not given regular interaction with other parrots. With greater interdisciplinary collaboration and training, we anticipate continued application of animal cognition research findings outside of the laboratory to improve how we as humans interact and work with animals.

HUMAN-INDUCED RAPID ENVIRONMENTAL CHANGE

Humans have changed the planet in the last few hundred years into something that is almost unrecognizable. As we come to terms with anthropogenic changes like global warming, pollution, and deforestation, we tend to focus on two things: how it impacts the environment and how that will impact us. Human-induced rapid environmental change (HIREC; Sih, Ferrari, & Harris, 2011) is also impacting animals—forcing them to either keep up with us by adapting or perish. Millions of years of evolution have created animals that are ideally suited to their environment, but for many animals, their environment has changed too quickly for evolution to keep up.

As we covered in Chapter 6, Cognitive Flexibility in Animals, animals that live in dynamic environments tend to be more cognitively flexible (Bond, Kamil, & Balda, 2007). For species whose behavior is driven by consistent, reliable cues in their environment, HIREC has rendered some of those cues unreliable, maladaptive, or both. For example, many male Australian jewel beetles (*Julodimorpha bakewelli*) die without ever successfully mating because they're fooled into "mating" with discarded shiny brown beer bottles that send their reproductive behavior into overdrive (Gwynne & Rentz, 1983; Schlaepfer, Runge, & Sherman, 2002). Sea turtles choke on plastic bags because, for them, a clear floating sac has always been a jellyfish—until recently. In these two cases, humans have unknowingly created a threat to survival by presenting stimuli that are dangerous mimics of what would otherwise be a normal part of the animal's natural history (Griffin, Tebbich, & Bugnyar, 2017).

Not all animals seem to be so negatively impacted by HIREC. Indeed, some have responded flexibly. For example, wildlife researchers have noticed a variety of changes in response to global warming, including shifting dietary preferences, exploiting new habitats, and altering features of reproduction (Beever et al., 2017). While these spontaneous changes do mitigate some of the stress caused by humans' behavior, the authors were quick to note that positive human intervention for the sake of conservation is still needed.

In light of HIREC, researchers agree that one of the next big shifts for animal cognition research will be to study cognition in action, and specifically as it relates to anthropogenic change and its effects on cognition (Griffin et al., 2017; Healy & Rowe, 2014). HIREC is the epitome of a dynamic environment. Determining which animals survive, or even thrive, among the chaos and change of human society, and how living with us has changed them are exciting research opportunities to explore. The escalating cognitive arms race between the local raccoons, my (ECW) neighbors, and their (now) bungee-cord fastened garbage cans would be a great case study. Unfortunately, someone already beat me to it. Empirical research on urban (i.e., lots of humans and their technology) versus rural (i.e., minimal human presence) raccoons has already shown that raccoons' cognitive abilities have changed as a result of indirect experience with our "natural habitat." Specifically, urban raccoons were more likely to figure out how to open a novel garbage can that had been suspended from a tree by a rope, they tried more strategies to open it (perhaps reflecting a flexible tool kit of methods drawn from the diversity of garbage cans in the world), and they persisted in working on opening it (as described in MacDonald & Ritvo, 2016).

By moving animal cognition further out of the laboratory, and realizing it's unrealistic to assume "the field" is untouched by humans, we can update the way that we ask research questions to reflect our rapidly changing world. Realistically, humans are not going anywhere anytime soon, and we both intentionally and unintentionally affect the behavioral and cognitive lives of animals. In the decades ahead, just how much we do should be a major area of study.

FINAL REMARKS

If there's one take-home message we hope to leave you with it's one of humility. For a long time, feelings of human superiority prevented us from learning about other species for the sake of learning about them. This rat cannot do calculus, so it *must* be dumber than me. That earthworm never once created a tool, so it *must* be inferior to humans. Though calls to abandon the human superiority perspective have been around for decades (e.g., Hodos & Campbell, 1969), it appears the pendulum has really swung away from the egotistical perspective that humans are the gold-star standard for all things cognitive. Instead, we hope that we have convinced you that cognition should be viewed as species-specific and tied to specific evolutionary needs (Premack, 2007).

You might have noticed that throughout this book, we limited the number of direct comparisons we made between humans' and animals' cognitive abilities. Human examples gave something familiar to anchor to when we broke down difficult concepts, but we did our best to avoid sizing up an animal's abilities to that of a human's. This was an intentional effort to replace the uninformed, human-centered comparative approach with a more contemporary perspective that each species is uniquely suited to deal with the ecological pressures of its environment. Rats do not do calculus, and worms do not use tools, because they do not need these abilities in order to successfully navigate their respective worlds. On the other hand, being able to see the world in ultraviolet is really helpful to rats, as is body regeneration for worms—two abilities we humans unfortunately do not have.

Whereas early pioneers like Washburn and Romanes explored a variety of species in their writing, research findings relevant to stereotypically intelligent animals like dolphins as well as dogs (talk about human bias!) dominate many of the social media videos, animal documentaries, and popular press articles. A taxonomic bias has been observed in higher education as well. In 2018, two entomologists surveyed 88 introductory biology textbooks published between 1907 and 2016 and made two very disturbing discoveries: (1) a 75% decrease in content about insects in recent (i.e., post-2000) textbooks compared to pre-1960 textbooks and (2) increasingly neutral language used to describe insects (Gangwani & Landin, 2018). The increase in neutral language might seem like a good thing, since positive language like *friend*, *amazingly*, and *help* anthropomorphizes insects and negative language like *pest*, *destructive*, and *disease transmitted* vilifies them. However, neutral language paints insects into the background scenery—no different from a rock, stream, or grass. Gangwani and Landin worry that neutral treatment, coupled with declining representation, could worsen humans' already minimal interest in scientific knowledge about insects and, by our estimation, other animals that are lesser known or valued such as spiders, fish, and reptiles.

We were curious to know whether published animal cognition literature has also decreased in its representation of species that are not big, cuddly, or assumed to be like us. So, we did some research! The *Journal of Comparative Psychology* has been around almost as long as animal cognition itself, with the first volume being published in 1921. In increments of 20-year spans, we counted the number of published primary research articles that used reptiles, amphibians, insects, spiders, or fish as their research subjects. To our surprise, representation of these species increased from a combined three to a whopping 37. Furthermore, the word *invertebrate* appears in the title or subject in five articles from this journal in the past 20 years, compared to zero at the journal's inception. The greatest amount of biodiversity in research subjects occurred in the last 20-year span, from 1998 to 2018, signaling the opposite of what entomologists Gangwani and Landin (2018) noticed. We hope that our book has captured your attention and interest about the vast biodiversity we share the planet with—not just the charismatic megafauna we're used to seeing or reading about. Who knows, as we learn more about the minds of bees, fish, and frogs, we may become more empathetic and work harder and with greater urgency to protect our forests and waterways. By presenting some of our favorite examples of advanced cognition throughout the animal world, we hope that we've created space for you to think not just about animal cognition but also about our place within the animal kingdom. Because after all, we're animals, too!

REFERENCES

AAALAC International's Reference Resources. (n.d.). Retrieved December 17, 2018, from https://www.aaalac.org

Abramson, C. I. (2015). A crisis in comparative psychology: Where have all the undergraduates gone? *Frontiers in Psychology, 6*, 1500. doi:10.3389/fpsyg.2015.01500

Agnew, Z. K., Bhakoo, K. K., & Puri, B. K. (2007). The human mirror system: A motor resonance theory of mind-reading. *Brain Research Reviews, 54*(2), 286–293. doi:10.1016/j.brainresrev.2007.04.003

Beever, E. A., Hall, L. E., Varner, J., Loosen, A. E., Dunham, J. B., Gahl, M. K., … Lawler, J. J. (2017). Behavioral flexibility as a mechanism for coping with climate change. *Frontiers in Ecology and the Environment, 15*(6), 299–308. doi:10.1002/fee.1502

Berns, G. S., Brooks, A. M., & Spivak, M. (2015). Scent of the familiar: An fMRI study of canine brain responses to familiar and unfamiliar human and dog odors. *Behavioural Processes, 110*, 37–46. doi:10.1016/j.beproc.2014.02.011

Bond, A. B., Kamil, A. C., & Balda, R. P. (2007). Serial reversal learning and the evolution of behavioral flexibility in three species of North American corvids (*Gymnorhinus cyanocephalus, Nucifraga columbiana, Aphelocoma californica*). *Journal of Comparative Psychology, 121*, 372–379. doi:10.1037/0735-7036.121.4.372

Breland, K., & Breland, M. (1961). The misbehavior of organisms. *American Psychologist, 16*(11), 681–684. doi:10.1037/h0040090

Cammaerts, M. C., & Cammaerts, R. (2015). Are ants (*Hymenoptera, Formicidae*) capable of self-recognition? *Journal of Science, 5*, 521–532.

Gangwani, K., & Landin, J. (2018). The decline of insect representation in biology textbooks over time. *American Entomologist, 64*, 252–257. doi:10.1093/ae/tmy064

Griffin, A. S., Tebbich, S., & Bugnyar, T. (2017). Animal cognition in a human-dominated world. *Animal Cognition, 20*, 1–6. doi:10.1007/s10071-016-1051-9

Gwynne, D. T., & Rentz, D. C. F. (1983). Beetles on the bottle: Male buprestids mistake stubbies for females (*Coleoptera*). *Australian Journal of Entomology, 22*(1), 79–80. doi:10.1111/j.1440-6055.1983.tb01846.x

Healy, S. D., & Rowe, C. (2014). Animal cognition in the wild. *Behavioural Processes, 109*, 101–102. doi:10.1016/j.beproc.2014.11.013

Hodos, W., & Campbell, C. B. G. (1969). Scala naturae: Why there is no theory in comparative psychology. *Psychological Review, 76*, 337–350. doi:10.1037/h0027523

Kulick, D. (2017). Human-animal communication. *Annual Review of Anthropology, 25*, 357–378. doi:10.1146/annurev-anthro-102116-041723

MacDonald, S. E., & Ritvo, S. (2016). Comparative cognition outside the laboratory. *Comparative Cognition & Behavior Reviews, 11*, 49–62. doi:10.3819/ccbr.2016.110003

Premack, D. (2007). Human and animal cognition: Continuity and discontinuity. *Proceedings of the National Academy of Sciences, 104*, 13861–13867. doi:10.1073/pnas.0706147104

Schlaepfer, M. A., Runge, C., & Sherman, P. W. (2002). Ecological and evolutionary traps. *Trends in Ecology and Evolution, 17*, 474–480. doi:10.1016/S0169-5347(02)02580-6

Shaw, R. C., & Schmelz, M. (2017). Cognitive test batteries in animal cognition research: Evaluating the past, present and future of comparative psychometrics. *Animal Cognition, 20*, 1003–1018. doi:10.1007/s10071-017-1135-1

Sih, A., Ferrari, M. C., & Harris, D. J. (2011). Evolution and behavioural responses to human-induced rapid environmental change. *Evolutionary Applications, 4*, 367–387. doi:10.1111/j.1752-4571.2010.00166.x

Index

AAALAC. *See* Association for Assessment and Accreditation of Laboratory Animal Care
AALAS. *See* American Association for Laboratory Animal Science
affective empathy, 148
African elephants (*Loxodonta africana*), 219
African grey parrot (*Psittacus erithacus*), 36
  Alex, 38, 45, 108, 114, 232–235
  Cosmo, 42, 114–115
  intelligence of, 232–235
  language learning, 114–115
  referential signaling of, 108
  self-awareness of, 73
  use of social cues by, 116
Akeakamai (bottlenose dolphin), 110, 122–123
alarm calls, 104, 107, 109
Alex (African grey parrot), 38, 45, 108, 114, 232–235
altruism, and empathy, 150
American Association for Laboratory Animal Science (AALAS), 34, 35
American Sign Language (ASL), 108, 112
amoebas (*Amoeba* spp.), 184
anecdotes, 33, 50
animal-centered anthropomorphism, 49
animal culture, 156–161
  birds, 159–160
  culture, definition of, 156
  fish, 160
  genetic testing, 159
  group-level differences within species, 158–159
  Imo (Japanese macaque), 161–164
  insects, 160–161
Animal Intelligence (Romanes), 5–6, 8, 50
animal law, 247
Animal Mind, The (Washburn), 11–12, 250
animal models, 21–25
  brain models, 23–24
  evaluation of, 24–25
  medical models, 22–23
anthropodenial, 16
anthropomorphism, 12–15, 48–49, 50
  animal-centered, 49
  human-centered, 49
  and mirror mark test, 73
ants (*Pogonomyrmex californicus*)
  communication between, 118, 119
  self-awareness of, 73
  tandem running, 144

INDEX

anxiety/anxiousness, 83, 84–85
apes. *See also* chimpanzees; gorillas (*Gorilla gorilla*); orangutans
  communication, 115–116, 123–124
  learning from others, 143
  use of gestures by, 115–116
appropriateness, and creativity, 194, 197, 203
aquariums, 193. *See also* zoos
  enriched environments in, 89, 193
  research in, 44
archer fish (*Toxotes jaculatrix*), 186
Aristotle, 21
ASD. *See* autism spectrum disorder
Asian elephants (*Elephas maximus*), 72
ASL. *See* American Sign Language
Association for Assessment and Accreditation of Laboratory Animal Care (AAALAC), 251
Association of Zoos and Aquariums (AZA), 88, 193
auditory discrimination task, 78–79
Austin (chimpanzee), 113
autism spectrum disorder (ASD)
  people with, TAGteach, 166
  and theory of mind, 91–92
axolotls (*Ambystoma mexicanum*), 22–23
AZA. *See* Association of Zoos and Aquariums

baboons (*Papio* spp.), 222
Bailey, Marian Breland, 180–181
Bandura, Albert, 142, 232
Barnum Effect, 238

bat (*Phyllostomus hastatus*), 141
BCCs. *See* behavioral correlates of consciousness
bees. *See* honeybees
behavioral correlates of consciousness (BCCs), 69, 71
  metacognition, 78–81
  self-awareness, 71–74
  theory of mind, 74–77
behaviorism, 6–7, 13, 51, 142, 250
belugas (*Delphinapterus leucas*), 158
Belyayev, Dmitry, 117
Bentham, Jeremy, 69
Betty (New Caledonian crow), 197–200
bias, human superiority, 12, 17
Big Five personality traits, 220, 221, 224, 236
blindsight, 69
blue jays, 199
bonobos (*Pan paniscus*)
  Kanzi, 17–21, 113–114
  personality of, 224
bottlenose dolphins (*Tursiops truncatus*)
  Akeakamai, 110, 122–123
  communication, 110, 116, 120, 122–123
  emotions in, 85
  Phoenix, 110, 122–123
  self-awareness of, 72
  and syntactic commands, 110
  use of social cues by, 116
bowerbird (*Ptilonorhynchidae* spp.), 197
brain
  and consciousness, 70
  development, and language learning, 126
  scanning, 249

size
  and play, 191
  in social species, 143
  and social interactions, 147–148
  Urbach–Wiethe disease, 81
brain models, 23–24
  fruit fly, 23–24
  pond snail, 24
Breland, Keller, 180–181
bridge (operant conditioning), 165–166
brown rat. *See* Norway rat (*Rattus norvegicus*)
bumblebees (*Bombus terrestris*), 143
  culturally transmitted behaviors, 160
  problem solving by, 186–187
Burghardt, Gordon, 14

C-BARQ. *See* Canine Behavioral Assessment and Research Questionnaire
C-Well. *See* Cetacean Welfare Assessment
Campbell's monkeys (*Cercopithecus campbelli*), 109
canaries, 125
Canine Behavioral Assessment and Research Questionnaire (C-BARQ), 225
capuchin monkeys (*Sapajus apella*)
  contagious behaviors in, 144–145
  imitation in, 147
  self-awareness of, 72
  social referencing in, 155
case study method, 37–39
CAT. *See* consensual assessment technique
cats, 34, 55, 154
Cattell, James McKeen, 10
Cattell–Horn–Carroll (CHC) theory, 229, 230
causation
  versus correlation, 40
  criteria for, 41
  and experiments, 42
Cetacean Welfare (C-Well) Assessment, 227–228
cetaceans, 119–120, 158
Chalmers, David, 82
chameleon effect, 147
CHC. *See* Cattell-Horn-Carroll theory
cheetah (*Acinonyx jubatus*), 148
child development
  lying, 74–75, 90
  theory of mind in, 90–92
chimpanzees (*Pan troglodytes*), 17–21
  Austin, 113
  culturally transmitted behaviors, 158
  emulation in, 146
  metacognition of, 79
  Nim Chimpsky, 112
  personality of, 224
  problem solving by, 185
  referential signaling of, 108
  Sherman, 113
  and social cues, 117
  social referencing in, 154
  teaching language to, 112, 113
  theory of mind of, 75–76
  use of gestures by, 115, 116
  Washoe, 108, 112
chipmunk (*Marmotini* spp.), 46
Chomsky, Noam, 233–234
Christianity, 3, 17
cleaner wrasse (*Labroides dimidiatus*), 182–183
Clever Hans (horse), 52–55, 248

cockroaches, 9
cognitive bias task, 83–84, 85, 227
cognitive development, theory of (Piaget), 90–91
cognitive empathy, 148–149
cognitive enrichment, 88–89
cognitive flexibility, 177–178, 204–206
  Betty (New Caledonian crow), 197–200
  innovation, 193–197
  instincts, 178–181
  planning and forethought, 181–184
  play behavior, 189–193
  problem solving, 185–189
cognitive resources, 126, 193, 203
color
  communication about, 108
  learning about, 233
common brown rat (*Rattus norvegicus*), metacognition of, 79–80
communication between animals, 103–106
  dishonest signaling, 103–105
  features, 106–111
  gestures and social cues, 115–118
  insect communication, 118–119
  referential signaling, 107–109
  senses, 105
  syntax, 109–111
  teaching language to animals, 18–20, 111–115
comparative experiments, 42–43
conditioned learning, 106
confidence judgments, and metacognition, 79–80
conscientiousness, 221
consciousness
  awareness of others, 74–77
  behavioral correlates of, 69, 71
  defining, 67–68
  emotions, 81–86
  Happy (elephant), 86–89
  integrated information theory, 70
  metacognition, 78–81
  mirror mark test, 71–74, 86–88, 90–91, 92–94
  phenomenal, 68
  psychological, 68
  self-awareness, 71–74
  structure–function problem, 68–69, 70
consensual assessment technique (CAT), 201
consolation behavior, and empathy, 151
construct validity, 222–223
contagious behaviors, 144–145
contagious yawning, and empathy, 151
continuity of species, 4–5, 13, 16, 25
control conditions, 92–94
convergent evolution, 57, 58
Copernicus, 32
copying (learning), 144–147
correlation, 39–41
cortisol, 219
corvids
  culture, 159
  planning, 199
  theory of mind of, 77
  toolmaking, 198
Cosmo (African grey parrot), 42, 114–115
cows (*Bos Taurus*), 85
crayfish (*Procambarus clarkia*), emotions in, 84–85
creative metacognition, 203
creativity, 194–195, 196, 197–198, 200–204

Big-C creators, 201–202
little-c creators, 202
mini-c creators, 202–203
Person, 201–203
Press, 200
pro-c creators, 202
Process, 201
Product, 195, 200–201
critical anthropomorphism, 14
critical period hypothesis, 124–125, 126, 127
crocodiles (*Crocodylinae spp.*), 192
crow (*Corrus spp.*), 36
crystallized intelligence, 229
cuckoo (*Cuculus canorus*), instincts of, 180
cultural intelligence hypothesis, 231
culture. *See also* animal culture
cuttlefish (*Sepia apama*), dishonest signaling of, 105

daffodil cichlid (*Neolamprologus pulcher*), 73
dart frogs (*Dendrobatidae* spp.), 192
Darwin, Charles, 1–2, 13, 14, 17, 21, 22, 25, 55, 58, 71, 82, 86, 230
data, 33, 45–46
data collection, 45–47
deep homologies, 58
deer, threat response of, 177
dependent variables, 42
Descartes, René, 1–4, 17, 177
dialects, in communication, 111, 120, 121
Diana monkeys (*Cercopithecus diana*), referential signaling of, 107
dishonest signaling, 103–105
displacement, language, 111
division of labor, 141

"do as I do" method, 147
documentaries, 247
Dognition, 225
dogs
aggression of, 42
brain scanning, 249
Canine Behavioral Assessment and Research Questionnaire (C-BARQ), 225
empathy in, 150
and gestures/social cues, 117–118
learned helplessness of, 84
learning from other species, 143
metacognition of, 80
personality of, 225
predisposition to pointing sensitivity, 179
reliance on humans, and domestication, 42–43
self-awareness of, 72
social referencing in, 155
stress behavior, 46
theory of mind of, 76–77
dolphins, 218. *See also* bottlenose dolphins (*Tursiops truncatus*)
communication, 116–117, 119–124
culturally transmitted behaviors, 158
imitation in, 147
innovation of, 194–195
metacognition of, 78–79
personality of, 217–218, 220
planning, 182
self-awareness of, 72
sponging behavior of, 159
use of social cues by, 116–117
vocalization, 220
domestic cats (*Felis catus*), 148

domestication hypothesis, 117
dopamine, 23–24
Down syndrome, and theory of mind, 91–92

egocentrism, 91
electroencephalograph, 250
elephants (*Elephas maximus*), 54
   Happy, 86–89
   personality traits of, 219
   self-awareness of, 72
emerald anole (*Anolis evermanni*), problem solving by, 187
emotional contagion, 149–150
emotional empathy. *See* affective empathy
emotional well-being, 216, 226–228
emotions, 81–86, 218–220
empathy, 87, 148–153
   affective, 148
   and altruism, 150
   cognitive, 148–149
   and consolation behavior, 151
   and contagious yawning, 151
   definition of, 148
   emotional contagion, 149–150
   and helping behavior, 150–151
   and mirror neuron activation, 152
   and oxytocin release, 152
   and survival, 149
emulation, 146
encultured animals, 45
energy expenditure, and play, 191
environment
   adaptation, and play behavior, 193
   and behavior, 142, 178
   enrichment of, 89, 193
   human-induced rapid environmental change (HIREC), 252–253
   and instincts, 179
   and personality, 217
   research, 43–45
epigenetics, 178
ethics of animal research, 25, 34–35, 251
ethograms, 46–47, 59–60, 87, 129, 167, 168
evolution
   and personality traits, 225–226
   theory of, 4, 21, 55
evolutionary biology, 55–58
experiments, 41–42
   comparative, 42–43
extraction task, 186
extraversion, 220
extrinsic motivation, and creativity, 200

factual knowledge. *See* crystalized intelligence
false killer whales (*Pseudorca crassidens*), self-awareness of, 72
fear, 81
   and cortisol levels, 219
   of novel objects, 195, 196
   and social referencing, 154
feeding enrichment, 89
findings, interpretation of, 48–52
fireflies (*Photuris* spp.), dishonest signaling of, 104, 118
fish. *See also specific entries*
   culturally transmitted behaviors, 160
   emotions in, 83
   planning, 182–183
   problem solving by, 186
   self-awareness of, 73
five-factor model. *See* Big Five personality traits
fluid intelligence, 229

fMRI. *See* functional magnetic resonance imaging
food
   caching, by birds, 183, 190, 198–199
   and dishonest signaling, 104
Forer Effect. *See* Barnum Effect
forethought, 181–184
foxes (*Vulpes vulpes*), 6
   and gestures/social cues, 117
   play behavior of, 191–192
frogs
   dishonest signaling of, 104
   play behavior of, 192
fruit fly (*Drosophila melanogaster*), 23–24, 84
functional magnetic resonance imaging (fMRI), 249
fur rubbing, 145

Gallup, Gordon, 71, 72, 90
gazing (social cue), 116, 117
general intelligence (*g*), 229, 230
generativity, language, 111, 233–234
genetic testing, 159
genetics
   and emotions, 81
   epigenetics, 178
   and personality, 217
   phylogenetics, 56, 58
   and play behavior, 191
Genie, 125–128
gestures, 115–118, 120, 122, 123, 179
giant pandas (*Ailuropoda melanoleuca*), self-awareness of, 73
global warming, 252
goats (*Capra aegagrus hircus*), 117
Goodall, Jane, 17, 39, 158
gorillas (*Gorilla gorilla*)
   Koko, 108
   self-awareness of, 72
Gould, Stephen J., 4
great tit (*Parus minor*), communication of, 110
green frogs (*Rana clamitans*), 104
Griffin, Donald, 14
grizzly bears (*Ursus arctos*), problem solving by, 186, 188
group movement patterns of animals, 182
guinea pigs (*Cavia porcellus*), 22
guppies (*Poecilia reticulate*), 160

hamsters (*Mesocricetus auratus*), 46
Happy (elephant), 86–89
Harvey, William, 2, 3
hearing. *See* vocal communication
helping behavior, and empathy, 150–151
hermit crab (*Pagurus bernhardus*), 221
HIREC. *See* human-induced rapid environmental change
honeybees
   Apis mellifera, metacognition of, 80
   emotions in, 85
   olfactory communication, 118–119
   waggle dance, 118
horses (*Equus caballus*)
   Clever Hans, 52–55, 248
   learning, 38
   self-awareness of, 72
human-centered anthropomorphism, 49
human-induced rapid environmental change (HIREC), 252–253
human superiority, 3, 17
human uniqueness, 15–17

INDEX

Humane Society of the United States, 228
humpback whale (*Megaptera novaeangliae*)
  culturally transmitted behaviors, 158
  songs, 111, 158
hybrid research locations, 44–45
hyenas (*Crocuta crocuta*), 186

IACUC. *See* Institutional Animal Care and Use Committee
IIT. *See* integrated information theory
iguanas (*Iguana iguana*), 85
imitation, 146–147
Imo (Japanese macaque), 161–164, 194
independent variables, 42
individual differences between animals, 215–217, 238–240
  Alex (African grey parrot), 232–235
  bold-*vs.*-shy test, 238–240
  personality, 217–226
innovation, 193–197
insects. *See also specific entries*
  communication, 118–119
  culture, 160–161
  literature about, 254
  stress behavior, 84
insight, 185, 188–189
instinctive drift, 181
instincts, 178–181
Institutional Animal Care and Use Committee (IACUC), 34, 35
integrated information theory (IIT), 70
intelligence, 216, 228–231
  Cattell-Horn-Carroll (CHC) theory, 229, 230
  vs. cognition, 228–229
  crystalized intelligence, 229
  cultural intelligence hypothesis, 231
  fluid intelligence, 229
  general intelligence (g), 229, 230
  Primate Cognition Test Battery (PCTB), 231
  Triarchic Theory of Intelligence (TTI), 229
interdisciplinary collaborations, 250–251
interobserver reliability, 47–48, 224
introspection, 6
invasive species, innovation of, 196

jackdaws (*Corvus monedula*), self-awareness of, 73
Japanese macaque (*Macaca fuscata*)
  innovation of, 194
  play behavior of, 191
  social cognition of, 161–164
jewel beetles (*Julodimorpha bakewelli*), 252
jumping spiders, problem solving by, 187–188

Kanzi (bonobo), 17–21, 113–114
killer whales (*Orcinus orca*)
  communication, 120–121
  seal hunting technique of, 144
Köhler, Wolfgang, 185–186, 188
Koko (gorilla), 38, 45, 108

L-DOPA, 23
laboratory research, 44
Lana (chimpanzee), 113
language
  American Sign Language (ASL), 108, 112

critical period hypothesis, 124–125, 126, 127
generativity of, 111, 233–234
raising children without, 124–128
teaching, to animals, 18–20, 111–115, 122–124
Yerkish, 18–20, 113
law. *See* animal law
law of effect, 188
learned helplessness, 84
learning
copying, 144–147
and instincts, 180–181
language. *See under* language
observational, 143
from others, 142–148, 160
and snails, 24
social, 142, 143–144, 158–159, 165, 232
trial-and-error, 179, 188
vocal, 124
leopards (*Panthera uncia*), 44
Lewes, George Henry, 13
lexigrams, 18–20, 113
lizards
play behavior of, 192
problem solving by, 187
local enhancement, 145
location, research, 43–45
Lorenz, Konrad, 14
lying, and children, 74–75, 90

macaques
culture, 161–164
personality of, 219, 223–224
social referencing in, 154
machines, animals as, 3–4, 177
made-up languages, 18–20, 113
magazines, 247
magnetic resonance imaging (MRI), 249, 250

MAP. *See* modal action pattern
marmoset (*Callithrix jacchus*), 224, 226
mating, and dishonest signaling, 104–105
MBTI. *See* Myers–Briggs Type Indicator
medical models, 22–23
axolotls, 22–23
zebrafish, 23
meerkats (*Suricata suricatta*), 141, 226
memory
of elephants, 86
prospective, 181
and snails, 24
visual memory task, 43
mental illness, and creativity, 203
mental time travel, 181, 199
metacognition, 78–81
mind, 3, 13. *See also* theory of mind
mind–body dualism, 3
structure of, 6
Minnesota Multiphasic Personality Inventory, 236–237
minnows (*Phoxinus phoxinus*), 160
mirror image. *See* mirror mark test
mirror mark test, 71–74, 86–88, 90–91, 92–94
mirror neuron activation, 152, 249–250
MMPI. *See* Minnesota Multiphasic Personality Inventory
modal action pattern (MAP), 144
model-rival technique, 114, 232, 234
monarch butterfly (*Danaus plexippus*), 103

monkeys. *See also specific entries*
  communication, 109
  empathy in, 149
  metacognition of, 79
  self-awareness of, 72
Morgan, C. Lloyd, 50, 51
Morgan's Canon, 51–52, 54, 74, 188
MRI. *See* magnetic resonance imaging
Myers–Briggs Type Indicator (MBTI), 237
myna (*Acridotheres tristis*), 221–222

Nagel, Thomas, 69, 70
naked mole rat (*Heterocephalus glaber*), 23
Native Language Magnet Theory, 126
natural habitat, research in, 44
natural history, 5, 251
natural parks, research in, 44
naturalistic observation, 36–37, 44, 82
naturalists, 5–6
negative correlation, 40
negative punishment, 165
negative reinforcement, 165
negative stress, 227
neocortex size, and primate group size, 39–40
neophilia, 195–196, 221
neuroscience technologies, 249–250
New Caledonian crow (*Corvus moneduloides*)
  Betty, 197–200
  self-awareness of, 73
  toolmaking, 197–198
Nim Chimpsky (chimpanzee), 112
Norway rat (*Rattus norvegicus*), 196–197
nose scrunching, 82

novelty, 239
  and boldness, 222, 239–240
  and creativity, 194, 195–196, 197, 203

object-choice task, 154–155
object permanence, 44–45
observational learning, 143
observational research. *See* naturalistic observation
octopus (*Octopus vulgaris*), 143, 192
olfactory communication, 105, 118–119
open-field test, 221–223
openness to experience, 221
operant conditioning, 164–166
operational definitions, 33–34, 46, 59, 60, 224, 239
orangutans (*Pongo* spp.)
  dishonest signaling of, 104
  planning, 182
  Sandra, 251
orca (*Orcinus orea*), 120
oxytocin release, and empathy, 152

Pan/Homo culture, 18
Parkinson's disease (PD), 23–24
parrots (*Forpus paserinus*). *See also* African grey parrot (*Psittacus erithacus*)
  cognitive flexibility of, 180
  culture, 159–160
  intelligence of, 232–235
  language learning, 112, 114–115
  problem solving by, 188
  vocal learning of, 124
parsimony, 51
PCTB. *See* Primate Cognition Test Battery
PD. *See* Parkinson's disease
peacock (*Pavo cristatus*), 105
Pepperberg, Irene, 17, 232

person-situation controversy, 235–236, 237
personality, 215, 216, 217–226
  and behavior, 220–222
  emotional states, interpretation of, 218–219
  and environment, 217
  and evolution, 225–226
  and genetics, 217
  inventories, 236–238
    Minnesota Multiphasic Personality Inventory (MMPI), 236–237
    Myers-Briggs Type Indicator (MBTI), 237
  measurement of, 218
  models, 223, 224–225
    Canine Behavioral Assessment and Research Questionnaire (C-BARQ), 225
    preexisting, adapting, 224
  open-field test, 221–223
  person-situation controversy, 235–236, 237
  of primates, 223–224
  research, 218
    construct validity, 222–223
    reliability, 223, 224–225
    validity, 223, 225
  tests, 235–238
  theories, 236
    trait theory, 236
  and vocalization, 220
perspective taking, 75, 91–92
phenomenal consciousness, 68
pheromone trails, 119
phi, 70
Phoenix (bottlenose dolphin), 110, 122–123
phonemes, 126
phylogenetic trees, 56–57
phylogenetics, 56, 58

Piaget, Jean, 90–91
pigeon (*Columba livia*)
  metacognition of, 80
  problem solving by, 188–189
pigs (*Sus* spp.), 180
planning, 181–184
play behavior, 189–193
  and adaptation to changing environments, 193
  and brain size, 191
  criteria, 190
  and energy expenditure, 191
  functions of, 192
  and genetics, 191
  and habitat, 191
  and mating, 191
  and social life, 191
playback studies, 128–129
pointing (social cue), 116, 117, 123, 155, 179
pond snail (*Lymnaea stagnalis*), 24
popular press articles, 247
positive correlation, 40
positive punishment, 165
positive reinforcement, 89, 164–165
positive stress, 227
power studies, 39, 74
prairie dog (*Cynomys gunnisoni*), 111
Primate Cognition Test Battery (PCTB), 231
problem solving, 185–189, 197, 234. *See also* fluid intelligence
prospective memory, 181
psychological consciousness, 68
psychologists, 6–7
punishment, 164–165

qualitative data, 45–46
quantitative data, 46
quantities, perception of, 233

raccoons (*Procyon lotor*), 253
rats, 7
    emotions in, 84, 85
    empathy in, 149, 153
    innovation of, 196
    learning from other species, 143
    metacognition of, 79–80
    open field test, 223
ravens (*Corvus corax*)
    culture, 159–160
    theory of mind of, 77
referential signaling, 106, 107–109
reinforcement, 164–165
reliability
    assessment of, 47–48
    personality trait models, 223, 224–225
reptiles
    play behavior of, 192
    problem solving by, 187
research designs, 35–43
    case study method, 37–39
    comparative experiments, 42–43
    correlation, 39–41
    experiments, 41–42
    naturalistic observation, 36–37
research findings. *See* findings, interpretation of
research locations, 43–45
research question, 33–34
    and naturalistic observation, 36–37
rhesus macaques (*Macaca mulatta*), personality of, 219, 223–224
ring-tailed lemurs (*Lemur catta*), 178
risk taking, and innovation, 195–196
Romanes, George, 5–6, 8, 13, 50, 254
rooks, 198

Sally-Anne test, 91–92
Sandra (orangutan), 251
sankofa bird, 1
Savage-Rumbaugh, Sue, 18–20, 21
science, 32–33
scientific method, 32, 33, 48
Scientific Revolution, 2
scorpions, 6
sea turtles, 252
segmentation of the language stream, 126
self, 71, 78, 86–87, 90
self-awareness, 71–74, 87, 92. *See also* mirror mark test
self-concept, 216
sheep (*Ovis aries*), 84
Sherman (chimpanzee), 113
shrimps (*Synalpheus regalis*), 141
sight, and communication, 105
signature whistles, of dolphins, 121
Skinner, B. F., 142, 165
smell. *See* olfactory communication
snails, 24
snow leopards, 44, 45
social cognition, 141–142, 167–168
    animal culture, 156–161
    empathy, 148–153
    Imo (Japanese macaque), 161–164
    learning from others, 142–148
    social referencing, 153–156
social consciousness, 11
social cues, 115–118, 123, 143, 153–156
social desirability bias, 220
social intelligence hypothesis, 143
social learning theory, 142, 143–144, 158–159, 165, 232
social media, 247

social referencing, 153–156
songbirds
　age-limited learners, 125
　communication, 110, 125
　critical period for, 125
　culture, 159, 160
　open-ended song learners, 125
soul, 3
South African fur seals (*Arctocephalus pusillus*), use of social cues by, 116
Spearman, Charles, 229
spiders (*Araneae spp.*)
　problem solving by, 187–188
　webs, and prey capture, 40
sponging behavior, of dolphins, 159
stem cells, 22
Sternberg, Robert, 229
stimulus enhancement, 145
stone flaking, by apes, 20–21
stress
　coping style, 227
　and emotional well-being, 227
　hormones, and emotional states, 219
　negative, 227
　positive, 227
　reactivity, 227
string-pull task, 186, 188, 204–206, 234
structuralism, 6
structured observation, 36, 59–60
syntax, communication, 106, 109–111

tadpoles, play behavior of, 192
TAGteach (Teaching with Acoustical Guidance), 166
tail flicking, 83
tameness, 191–192
tandem running, in ants, 144
taste, and communication, 105

technological advancements, 248–250
theory of mind, 74–77, 155
　and autism spectrum disorder, 91–92
　in child development, 90–92
　perspective taking, 75, 91–92
thigmotaxis, 221
Thorndike, Edward, 188
three mountain problem, 91
Titchener, Edward, 6
Tolman, Edward, 7
tone discrimination task, 84
Tonkean macaque (*Macaca tonkeana*), play behavior of, 191
toolmaking, 55–58, 104, 196, 197–198, 199
touch, and communication, 105
toys, zoo, 89, 193
trait theory, 236
transitive inference, 178
travel routes of animals, 182
trial-and-error learning, 179, 188
Triarchic Theory of Intelligence (TTI), 229
TTI. *See* Triarchic Theory of Intelligence
tufted capuchin monkeys, dishonest signaling of, 104
Turner, Charles Henry, 8–10
turtles (*Testudines spp.*), play behavior of, 192
tuskfish (*Choerodon schoenleinii*), innvoation by, 196

uncertain option task, 78–79
Urbach–Wiethe disease, 81

variables, 33
veined octopuses (*Amphioctopus marginatus*), 184

vervets monkeys (*Chlorocebus aethiops*), 158–159
viceroy butterfly (*Limenitis archippus*), 103
vivisection, 3
vocal learning, 124
vocal mimicry, 123
vocalization, 106, 107, 109–110, 114–115, 120–121, 220
Voltaire, 3–4, 5
von Frisch, Karl, 118

waggle dance, of honeybees, 118
Washburn, Margaret Floy, 8, 10–11, 250, 254
   Animal Mind, The, 11–12, 250
   on social consciousness, 11
Washoe (chimpanzee), 108, 112
wasps, 10
Watson, John, 6–7, 13
western scrub jays (*Aphelocoma californica*)
   metacognition of, 80
   planning, 183
   theory of mind of, 77
whales
   communication, 120–121
   seal hunting technique of, 144
   self-awareness of, 72
   songs, 111, 158
whistles, of dolphins, 120, 121, 123
wolves (*Canis lupus*), 117

yawning, contagious, 151
Yerkish (language), 18–20, 113

zebrafish (*Danio rerio*), 23
zoos, 193
   cognitive enrichment, 88–89
   enriched environment, 89, 193
   feeding enrichment, 89
   research in, 44
   toys, 89, 193
   welfare, 88–89

www.ingramcontent.com/pod-product-compliance
Ingram Content Group UK Ltd.
Pitfield, Milton Keynes, MK11 3LW, UK
UKHW051122220326
4879IPUK00012B/36